# Pergola

### The Club at NINE BRIDGES

# 클럽나인브릿지 파고라
## 자연을 닮은 건축과 기술

클럽나인브릿지 파고라: 자연을 닮은 건축과 기술
Pergola of The Club at NINE BRIDGES:
Architecture and Technology Resembling the Nature

초판 1쇄 인쇄 2018년 4월 10일 초판 1쇄 발행 2018년 4월 20일
First Printed 10 April, 2018 First Published 20 April, 2018

**지은이** 조호건축, CJ 건설, 일진유니스코 **발행인** 황용철 **편집총괄** 박성진 **편집** 공을채
**글** 이정훈, 황정서, 노민수, 피포 쵸라, 크리스티앙 프랑소아 **사진** 에프라인 멘데스(별도 표기 외)
**아트디렉터** 안지미 **디자인** 고인수 **국문교정교열** 하명란 **번역** 이단비, 홍근호
**영문감수** 나탈리 페리스 **발행처** (주)CNB미디어 **출판등록** 1992. 8. 8. (제300-2005-000142호)
**주소** 03781 서울특별시 서대문구 연희로 52-20 **전화** 02-396-3359 **팩스** 02-396-7331
**전자우편** editorial@spacem.org **홈페이지** http://www.vmspace.com
ISBN 979-11-87071-17-4(93540)

**Authors** JOHO Architecture, CJ E&C, ILJIN Unisco **Publisher** Hwang Yongchul
**Editor-in-Chief** Park Sungjin **Editor** Kong Eulchae **Text** Lee JeongHoon, Hwang Jeongseo, No Minsu,
Pippo Ciorra, Christian FRANÇOIS **Photographer** Efaín Méndez (unless otherwise indicated)
**Art Director** AN Jimi **Design** Ko Insu **Korea language Proofreader** Ha Myungran
**Translator** Lee Danbi, Hong Keunho **English language Editor** Natalie Ferris
**Publishing** SPACE BOOKS, an imprint of CNB media **Registration** 1992. 8. 8. (300-2005-000142)
**Address** 52-20, Yeonhui-ro, Seodaemun-gu, Seoul, Korea 03781 **Tel** +82-2-396-3359
**Fax** +82-2-396-7331 **E-Mail** editorial@spacem.org **Homepage** http://www.vmspace.com
©JOHO Architecture, CJ E&C, ILJIN Unisco 2018. Printed in Seoul, Korea

* 파본이나 잘못된 책은 구입처에서 바꾸어 드립니다.
* 이 책은 저작권법에 따라 보호받는 저작물이므로 무단전재와 무단복제를 금지하며, 이 책 내용의 일부 또는 전부를
  이용하려면 반드시 사전에 저작권자와 출판권자의 서면 동의를 받아야 합니다.
* 이 책에 수록된 사진 중 저작권자를 확인할 수 없는 일부는 저작원을 명기하지 못했습니다. 확인하는 대로 저작권자를
  명기하고 출판 동의를 구하겠습니다.
* 이 도서의 국립중앙도서관 출판예정도서목록(CIP)은 서지정보유통지원시스템 홈페이지(http://seoji.nl.go.kr)와
  국가자료공동목록시스템(http://www.nl.go.kr/kolisnet)에서 이용하실 수 있습니다.(CIP제어번호: CIP2018010239)

* All rights reserved. No part of this publication may be reproduced, stored in a retrieval system, or transmitted in any
  form or by any means, electronic, mechanical, photocopying, recording, or otherwise, without prior consent of the
  publisher.
* All efforts have been undertaken to duly identify the copyright holders as well as all contributors any error or omissions
  that may have occurred will be corrected in future editions where applicable.

**Pergola
of The Club
at NINE BRIDGES**

**Architecture and
Technology Resembling
the Nature**

JOHO Architecture,
CJ E&C, ILJIN Unisco

SPACE
BOOKS

**클럽나인브릿지 파고라: 자연을 닮은 건축과 기술**

**서문** 이정훈 —— 006
**자연과 건축을 연결하다** 이정훈 —— 008
**호흡하는 파빌리온** 피포 쵸라 —— 038
**철과 유리의 건축 역사** 크리스티앙 프랑소아 —— 054
**포토 에세이** 에프라인 멘데스 —— 072
**디지털을 통한 건축구조와 혁신** 일진유니스코 —— 128
**현장의 기록** CJ건설 —— 162

# Contents

**Pergola of The Club at NINE BRIDGES:**
**Architecture and Technology Resembling the Nature**

006 —— **Prologue**   Lee JeongHoon

008 —— **Connecting Nature with Architecture**   Lee JeongHoon

038 —— **The Pavilion that Breathes**   Pippo Ciorra

054 —— **History of Architecture with Steel and Glass**   Christian FRANÇOIS

072 —— **Photo Essay**   Efraín Méndez

128 —— **Architectural Structure and Innovation through Digital**   ILJIN Unisco

162 —— **Field Note**   CJ E&C

## 건축이 진정으로 완성되는 순간

건축의 진정한 완성은 건물의 물리적인 완성이 아니라 그것의 의미와 가치를
자료화하고 되새김질하며 다른 지역과 그것을 공유하는 것이라고 생각한다. 이러한
일은 한 권의 책이 과정을 기록하는 단순한 자료가 아니라 새로운 진보와 담론을
위한 발전의 토대이기 때문이다. 클럽나인브릿지 파고라의 출판은 그러한 점에서
건축가, 건축주, 시공자, 출판사, 사진작가 등이 모여 함께 그 과정과 의미를 기록한
노력의 산물이다. 사실 이번 출판은 현장에서 생경하게 느꼈던 제작 과정의 숭고함을
보다 보편적인 방식으로 공유하고 싶은 건축가의 열망에서 시작되었다. 이것은
비단 이중 덕트 시스템, 비정형 구조물과 유리 스킨의 기술적인 해결뿐 아니라 유리
건축의 역사에 있어서 이번 프로젝트가 갖는 의미를 파고든 것이다. 즉 독자들은
역사 속에서 이번 프로젝트의 맥락과 가치를 살피고, 기술적으로 어떻게 구현이
가능했는지 알아가는 즐거움이 있을 듯하다. 그런 측면에서 이 책은 단순한 기술서가
아니다. 새로운 관점을 바탕으로 자연과 건축의 관계를 유리 건축의 진화와 어떻게
맞물려가는지 설명하는 한편의 서사시와 같다.

과정을 기록하는 데 익숙지 않은 한국건축의 토양 속에서 이런 기록과 출판에
뜻을 같이해준 CJ 건설, 일진유니스코 관계자에게 깊은 감사를 전한다. 또한 출판을
진두지휘해준 공간서가와 어려운 여건 속에서 멋진 이미지들을 만들어준 사진작가
에프라인 멘데스에게도 고마움을 표하고 싶다.

이 책에서 크리틱과 역사적 아티클을 더해준 피포 쵸라 큐레이터와 크리스티앙
프랑소아 교수님에게도 감사를 드린다. 마지막으로 이 책이 많은 독자와 건축가들에게
큰 영감을 전해주길 기대해본다.

2018년 3월
이정훈 / 조호건축사사무소 대표

### The Moment When Architecture is Truly Complete

I believe the true completion of architecture is not in the physical completion of a building. Instead, it is at the stage of recording and ruminating on the significance and value of architecture, sharing these aspects with the wider world. One book is not just a simple record of data documenting the process, but a foundation for the development that aims towards progress and new discourse. In that sense, the publication of Pergola of the Club at NINE BRIDGES bears the fruit of the efforts of the architects, clients, construction engineers, publisher and photographers. In fact, this publication was instigated by the architect's fervent desire to share the sublime nature of the unfamiliar, on-site manufacturing process in a more general way. This not only probes into a technical solution for the double-duct system, an irregularly shaped structural form, or the glass skin, but also the significance this project holds in the history of glass architecture. In that regard, this book is not just a simple description of the project, but a piece of epic poetry that explains how the relationship between nature and architecture engages with the evolution of glass architecture.

I would like to express my profound gratitude to the officials at CJ E&C and ILJIN Unisco for their efforts in this documentation and publication, rising from the mists of Korean architecture that is not very accustomed to keeping a record of an overall process. I would also like to thank the photographer Efraín Méndez, who produced wonderful images under difficult circumstances. Moreover, I am grateful for the curator Pippo Ciorra and Professor Christian FRANÇOIS, who added a critique and historical article to this book.

Lastly, I hope this book will bring great inspiration to many readers and architects.

March, 2018

Lee JeongHoon / Principal, JOHO Architecture

이정훈
조호건축사사무소 대표

## 자연과 건축을 연결하다
## Connecting Nature with Architecture

**Lee JeongHoon**
**Principal, JOHO Architecture**

**건축은 자연 속에서 많은 영감을 받지만,
궁극적으로 건축의 구축 논리는 자연과 건축이 다름을 보여준다.
클럽나인브릿지 파고라에서 건축가는 단순히 공간을
확장한 것이 아니라, 자연에서 영감을 받고
3차원을 느낄 수 있도록 디자인했다.**

/
/

나는 어렸을 적에 할아버지 댁에 갈 때마다 마을 어귀에서 사람들을 맞이하던 나무를 아직도 기억한다. 나에게 그 나무는 단순히 고목 이상으로 할아버지 댁에 대한 하나의 장소를 연상하는 기억의 시작점이 되었다. 고목 밑의 평상은 동네 어른들의 쉼터이자 아이들의 놀이터가 되었고 때로는 마을에서 벌어지는 행사의 무대가 되기도 했다. 현장답사를 가서 클럽나인브릿지의 고목을 본 순간 유년에 나를 맞이하던 나무가 떠올랐다. 또한 나는 클럽하우스의 기능적 확장과 더불어 어떠한 방식으로 프로젝트를 발전시켜야 하는지 직관적으로 알 수 있었다. 그것은 아마 건축가로서뿐만 아니라 토양에서 자란 모든 이들이 공감할 수 있는 자연에 대한 보편적 이해일 것이다.

이번 프로젝트는 클럽나인브릿지 골프 클럽하우스의 공간 확장과 더불어 고목의 편안함을 재구축하는 데 큰 목적이 있었다. 확장의 전제는 무작위적으로 공간을 팽창하는 것이 아니라 기존의 고목에 대한 세심한 고려와 자연 질서에 대한 배려였다. 그것은 개별적 공간의 크기 문제를 넘어서 사물에 대한 새로운 관점을 의미한다. 즉 자연을 정복의 대상으로 바라보는 것이 아닌 건축 공간과 공생을 이루는 대등한 가치로 바라보는 가능성을 의미한다. 또한 그러한 공생의 가치는 그 대지가 펼쳐온 역사와 문화적 기반에 근거한다. 궁극적으로 건축은 인간이 필요로 하는 공간을 만들기 위한 것이지만 그것이 놓이게 되는 논리와 구축의 방식은 그 사회가 지닌 자연에 대한 감성과 인문학적 철학을 기반으로 한다.

**Although architecture is greatly inspired by nature to be constructed, the logic of architecture construction indicates that the architecture is ultimately different from nature. Architects not only simply extended the spaces of the Pergola of The Cub at NINE BRIDGES, but also designed it as a place to experience three-dimensional.**

I still remember the tree that greeted visitors to my grandfather's town, which I visited frequently as a child. To me, that the tree was something more than just an old tree – it became the starting point for a memory that was connected to a certain place, my grandfather's house. The space under the tree was sometimes used as either a resting place for the elders or a playground for the children, and at times it was also used as a stage for a town event. Upon seeing the old tree of The Club at NINE BRIDGES during site visit, it reminded me of the tree that I so often met in my youth. This allowed me to know intuitively how the project should be developed, along with the more functional aspects of the expansion to the clubhouse. It is a universal understanding of nature with which everyone—not just the architects—who grew up close to these lands can probably empathise.

For this project, the greater aim was to reinvigorate the feeling of comfort that comes from an old tree along with the spatial expansion of the golf clubhouse at The Club at NINE BRIDGES. The spatial expansion was not conducted aimlessly, but under the premise of deep respect towards a natural order and with a careful consideration of the old tree. This goes beyond the problems presented by the sizes of individual spaces and points towards founding a new perspective of objects. In other words, it hints at the possibility of seeing nature not as an object of submission but as a being of equal worth that forms a symbiotic relationship with architectural space. Moreover, the value of such symbiosis finds its ground in the historical and cultural lineage of that land. Ultimately, while we turn to architecture to

## 건축의 장소성에 관하여

서양 건축에서 델피Delphi는 공간적 모태가 되는 장소이다. 델피는 파르나소스Parnassus 산을 등지고 암피사만Gulf of Amphissa을 감싸 안은 지세 속에 통합의 장을 만들어낸다. 그곳에서 아폴론Apollon 신은 세상의 중심 역할을 하며 주변에 작게 흩어진 도시국가의 다름을 새로운 장소성을 통해 융합하는 역할을 한다. 이처럼 신화와 결합한 장소는 파르나소스의 대자연이 지닌 기운을 극대화하며 이곳에 오는 순례자들에게 장소의 의미를 전달한다. 델피의 공간적 배치가 통합의 장소성에서 시작되었다면 클럽나인브릿지 파고라(이하 나인브릿지 파고라)의 건축은 오래된 고목을 중심으로 공간을 재편하고자 시작되었다. 600여 년은 족히 된 제주의 팽나무는 골프장이 건설되기 훨씬 전부터 그곳에 자리하고 있었고, 의식적으로나 무의식적으로나 그곳의 팽나무는 골프클럽 내의 건축적 배치와 구축 논리를 지배하는 일종의 장소성이 되었다. 즉 델피의 신전은 자연적 지세 속에서 신탁의 장소성을 드러냄에 목적이 있다면 나인브릿지 파고라의 고목은 건축과 자연의 새로운 장소성의 논리를 제시한다는 점에서 차이가 있다.

흥미로운 점은 나인브릿지 파고라의 고목이 한국의 전통적인 나무가 가진 마을 통합과 상징성에서 유사한 역할을 한다는 점이다. 클럽하우스의 확장된 공간인 나인브릿지 파고라는 이곳을 방문하는 사람들의 동선을 컨트리클럽으로 연결하는 역할을 하고 평상시에는 클럽하우스 레스토랑과 연계되어 사용되지만 필요하면 세미나 및 연회장으로 사용된다는 점이 특징이다. 즉 이 프로젝트에서 고목은 클럽하우스의 중심을 상징함과 동시에 다른 기능의 클럽 영역을 연결 짓는 역할을 수행하는 것이다. 또한 나인브릿지 파고라는 기존 클럽하우스 공간들을 수평적으로 연결 짓는 연회장임과 동시에 일종의 개방적 정원이다. 즉 통로이자 실내 정원인 온실의 개념으로 구축되었다. 이것은 시각적으로는 열린 구조이지만, 기능적으로는 내부 공간이어야만 하고 동시에 실내외로 확장된 자연 정원의 기능을 담고자 한

fig. 1 이번 프로젝트는 클럽나인브릿지 골프 클럽하우스의 공간 확장과 더불어 고목의 편안함을 재구축하는 데 큰 목적이 있었다.

For this project, the greater aim was to reinvigorate the feeling of comfort that comes from an old tree along with the spatial expansion of the golf clubhouse at The Club at NINE BRIDGES.

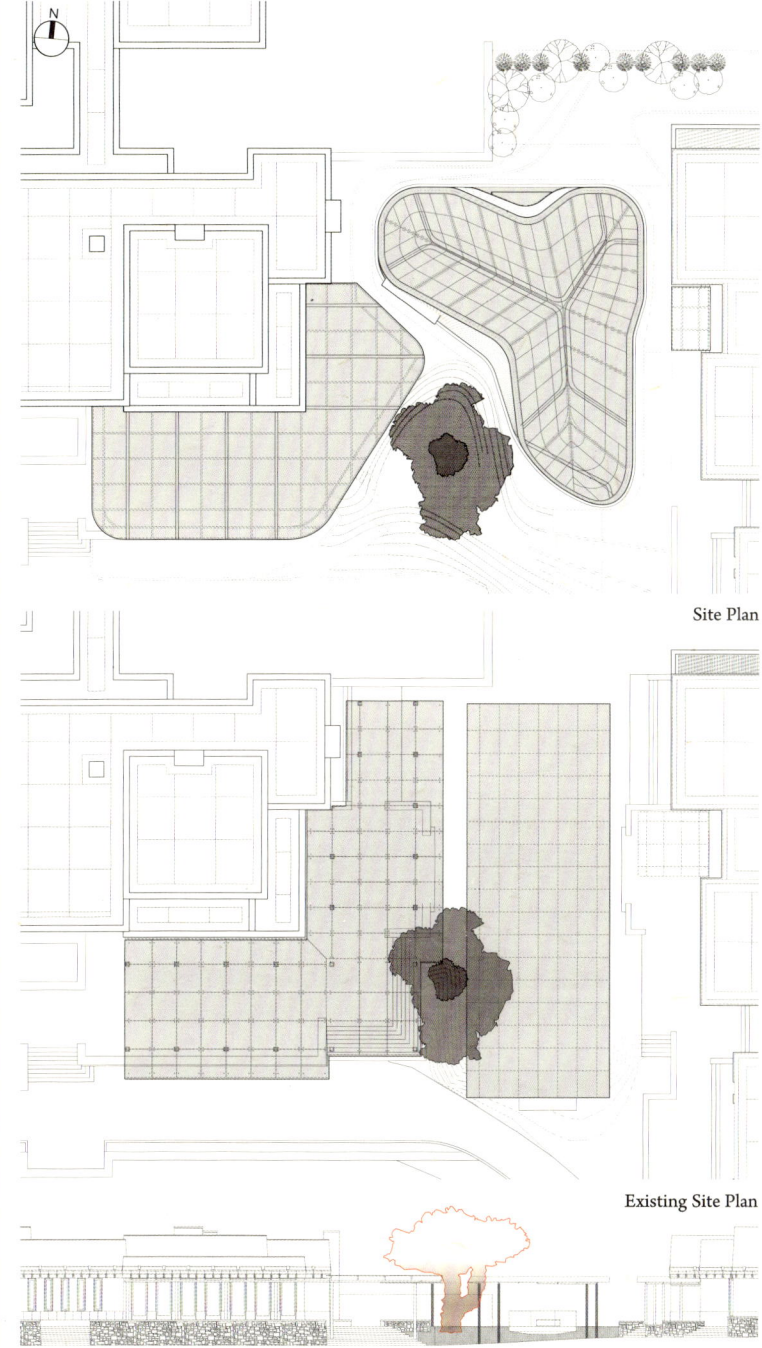

Site Plan

Existing Site Plan

fig. 1

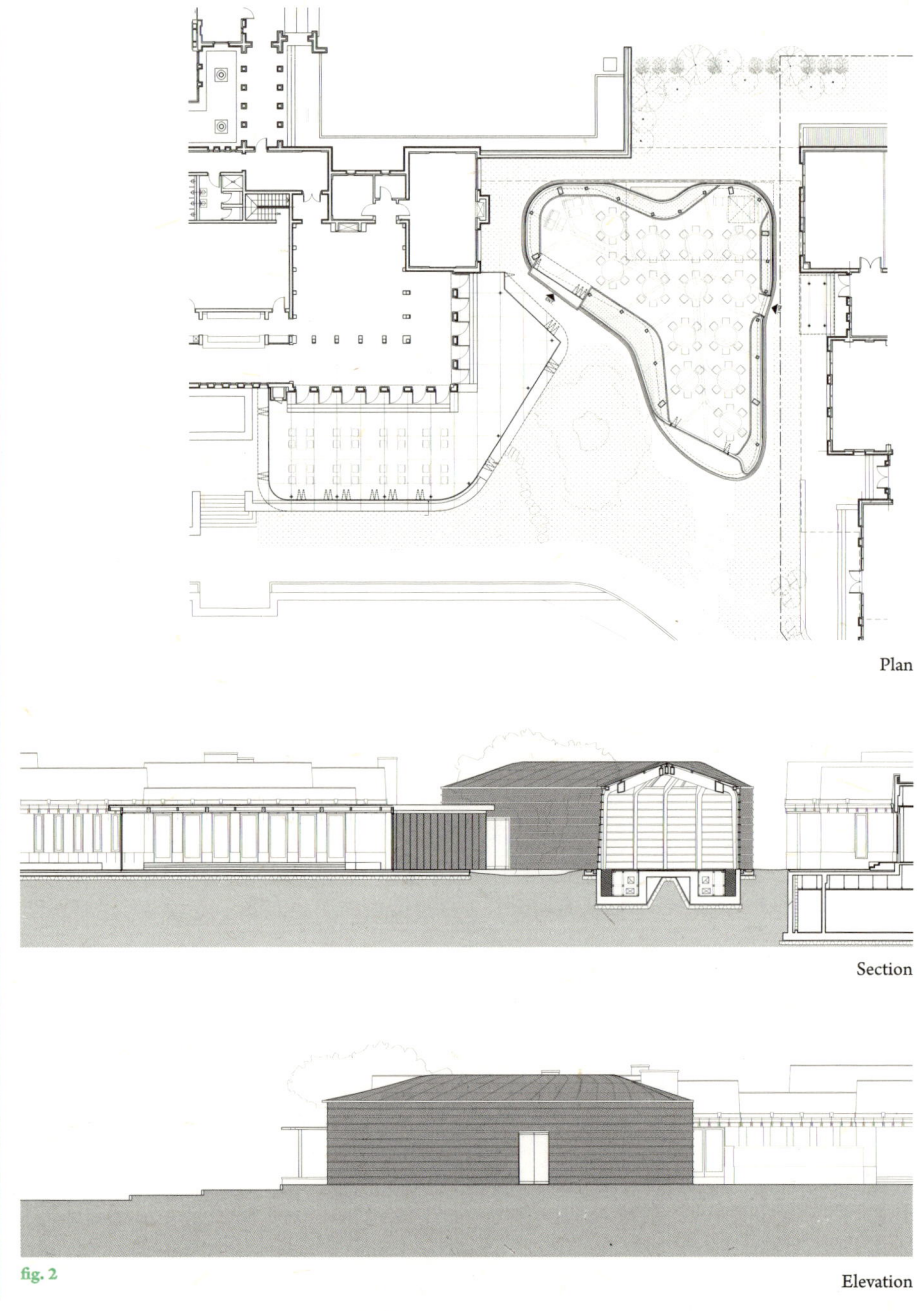

Plan

Section

Elevation

fig. 2

fig. 2 나인브릿지 파고라는 기존 클럽하우스 공간들을 수평적으로 연결 짓는 연회장임과 동시에 일종의 개방적 정원, 즉 통로이자 실내정원인 온실의 개념으로 구축되었다.

As a banquet hall that connects the original spaces of the clubhouse vertically, the NINE BRIDGES Pergola was also built also as an open garden with the concept of a greenhouse in mind, working as both a connecting path and an indoor garden.

create a space essential to human beings, its underlying logic and methods of construction are dependent on the emotions and humanitarian philosophy that society has adopted in view of nature.

### In Regard to the Locationality of Architecture

When it comes to western architecture, Delphi is a location that functions as a spatial matrix. As part of a topography that embraces the Gulf of Amphissa, with Mount Parnassus situated behind, Delphi forms a stage for integration. Functioning as the world's centre and as home to the god Apollo, it integrates the differences that have emerged between sparsely distributed neighbouring city-states by directing a new sense of locationality. This place enriched by mythology highlights the spiritual energy stored within the nature of Parnassus and delivers the meaning of location to the visiting pilgrims. If it can be said that the locational positioning of Delphi began from a locationality of integration, the Pergola of The Club at NINE BRIDGES (NINE BRIDGES Pergola) can be said to have begun with the intention of reorganising space with the old tree at its centre. This hackberry tree, which is at least 600-years old, stood in this place long before the construction of the golf course, and it has become a certain locational marker that dominates either consciously or unconsciously the architectural and constructional logic of the golf club. In other words, if the purpose of the shrine of Delphi is believed to be the revelation of locationality, as a divine oracle within the natural world, the old tree of the NINE BRIDGES Pergola differs in that it posits a new locational logic for both architecture and nature.

What is interesting is that the old tree of the NINE BRIDGES Pergola, like the trees of the Korean tradition, takes on an integrative and symbolic role representative of the town. The NINE BRIDGES Pergola, which is used as an expansion for the banquet hall of the clubhouse, is special in that while it is normally used as a connection to the country club

것이다. 이를 위하여 바닥에는 자연스럽게 제주산 현무암을 사용했고 내외부의 조경이 만나는 경계에는 제주에서 자라는 초화류 식물들을 심었다. 이는 나인브릿지 파고라가 대지가 지닌 자연적 풍토를 존중하면서 기존 지형에 순응하는 방식으로 배치되었음을 보여준다. 즉 제주의 풍토 위에 놓여진 투명 매스는 있는 그대로 자연을 품어 안으며 내외부 공간을 확장된 정원의 개념으로 연결 짓는다.

### 불확정 공간과 비물질성

나인브릿지 파고라는 법적으로 클럽하우스의 증축 공간이지만 역설적이게도 클럽하우스 체계 내에서 나인브릿지 파고라는 경계가 흐트러진 불확정 공간이다. 이곳은 프로그램들을 가능하게 하되 추후 다양한 용도로 활용될 수 있는 공간이다. 즉 기존 그리드 체계로 구축된 클럽하우스의 질서에 불확정 공간을 구성함으로써 기능적, 심리적으로 좀 더 유연하게 클럽나인브릿지를 재편한 것이다. 또한 나인브릿지 파고라는 그리드 방식으로 나누어진 기존 클럽하우스의 형태와 고목의 선형 사이에서 절묘한 완충제 역할을 수행한다. 나무와 골프장이 지닌 수려한 선형을 파고라 디자인의 모티프로 끌어안음으로써 클럽하우스와 골프장의 새로운 균형점을 형성하는 것이다.

유리는 이러한 점에서 공간적 균형을 매개하는 재료이다. 즉 기존의 클럽하우스와 고목 사이에 삽입된 투명한 유리 파빌리온을 통하여 자연과 골프장의 풍광을 내부로 끌어안는다. 유선형의 나인브릿지 파고라는 투명한 유리를 통해 기존의 질서에 자신의 존재를 최소한으로 드러내며 위장한다. 이러한 가운데 나무는 물질성과 비물질성의 대립과 조화의 축을 이루며 균형점을 찾는 새로운 장소성의 논리가 된다. 즉 이곳에서 투명한 물성은 존재의 최소화를 위한 전략이지만 역설적이게도 그곳의 풍광을 내부로 끌어안음으로써 기존의 강한 물질성으로 구축된 클럽하우스와 균형점을 찾은 것이다.

이러한 투명한 건축은 곧 물리적으로 최소한의 건축적 요소로 구성된 공간을 의미한다. 또한 건축적 구축의 문제는 곧 최소화된 설비와 일체화된 구조를 전제로 성립될 수 있다. 자신의 투명성을 드러내고자 함은 곧 자신을 형성하는 구축 논리를

and the clubhouse restaurant, it can also be used as a place for party and celebration. In other words, other than symbolising the centre of the clubhouse, the project is to use the old tree as a connection between the various sections and features of the club. Furthermore, as a banquet hall that connects the original spaces of the clubhouse vertically, the NINE BRIDGES Pergola was also built also as an open garden with the concept of a greenhouse in mind, working as both a connecting path and an indoor garden. While featuring a visually open structure, it had to be an indoor space in terms of its functions, and, simultaneously and necessarily, a natural garden that expands outdoors. For this purpose, Jeju basalt has been used in the flooring, and the local flora of Jeju has been planted at the intersection between the indoor and outdoor landscapes. This reveals how the NINE BRIDGES Pergola is positioned in a way that obeys the existing terrain while also especting the natural surrounding climate. The transparent mass that is placed within this Jeju landscape embraces the natural environment as it is, and thus connects the indoor and the outdoor spaces through the concept of a garden expansion.

### Indeterminate Space and Immateriality

While by definition it is an extension to the clubhouse, the NINE BRIDGES Pergola is also an indeterminate space of blurred demarcation within the overall clubhouse network of structures. What this means is that it is a space that could be utilized for a variety of purposes. By composing an indeterminate space within the systematic grid of the clubhouse, space becomes reorganized more flexibly in functional and psychological terms. Also, the NINE BRIDGES Pergola acts appropriately as a buffer between the grid-form of the original clubhouse and the organic linearity of the old tree. By embracing the elegant lines of the old tree and the golf course as its design motif, the Pergola forms a new balance point between the clubhouse and the golf course.

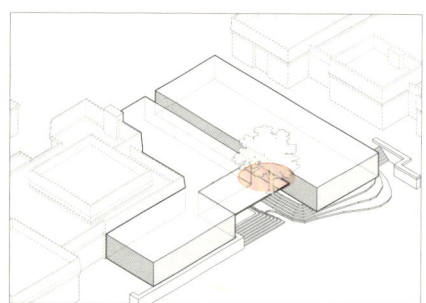

1. Existing Site

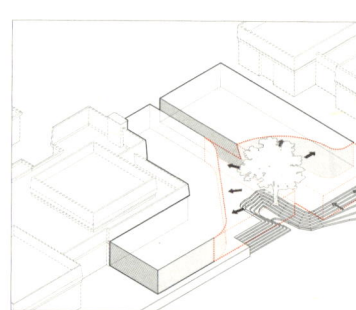

2. Introduction and Application of Natural Conditions

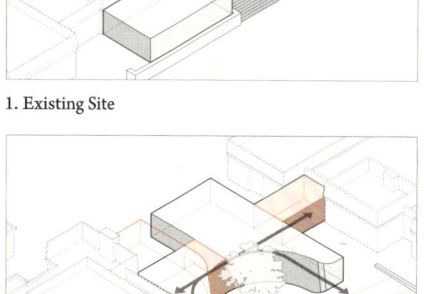

3. Adjusting the Landscape and Forms According to Programmes and Circulation

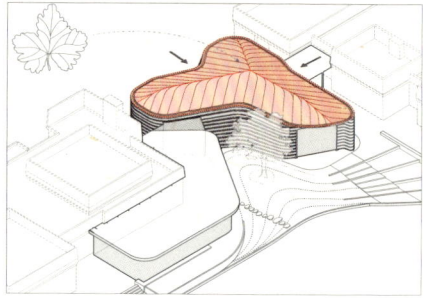

4. Adopting a Leaf Image that Looks Like Nature

1. Main Structure

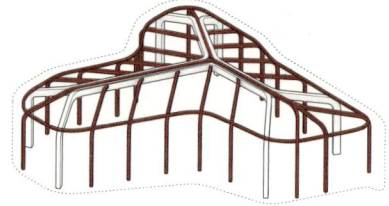

2. Addition of Sub Structure and Gutter

3. Addition of Horizontal and Vertical Façade Structure

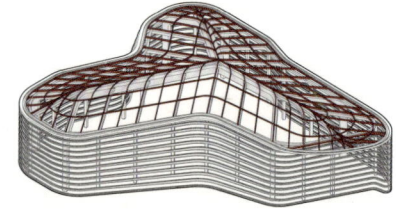

4. Addition of Roof Structure

fig. 3

fig. 3 나인브릿지 파고라는 자연에서 영감을 받고 이를 닮고자 시도한 프로젝트이다. 즉 투명한 스킨을 통하여 자연 속에 자신을 위장시킴과 동시에 외부로부터 내부공간을 보호한다.

The NINE BRIDGES Pergola is a project that not only borrows from nature, but one that strives to resemble it. Through its transparent skin, the project camouflages itself in nature, and simultaneously protects the inner space from the outside.

Glass is a material that mediates such spatial balance. Through the insertion of the transparent glass pavilion, between the original clubhouse and the old tree, the scenery presented by the nature and the golf course have been captured in the interior spaces. Because of the transparent property of glass, the streamlined NINE BRIDGES Pergola minimizes its impact on the existing order and disguises itself. Through this, the tree becomes the logic for a new locationality that seeks a balance between the axes of contrast and harmony between materiality and immateriality. As such, while this transparency is used strategically to minimise the structure's existence, pulling the scenery indoors, it also acts paradoxically as a serene counterpoint to the hard materiality of the clubhouse.

Architecture that employs transparency divines a space that is composed with the minimum architectural elements in physical terms. Such matters of construction in architecture can be put forward under the premise of minimal installation and a unified structure. This is because the desire to reveal oneself within transparency can be resolved by downplaying the structural logic that shapes oneself. What is interesting is that the project logic for the reconstruction of the locationality of an old tree moves beyond the materiality of transparency, and thereby the condensed construction logic arrives at a substantial problem upon installation and composition. In that sense, while the NINE BRIDGES Pergola began by questioning locationality—i.e., something very basic and fundamental to architecture—it is also a project that has undergone numerous technical evolutions in order to provide an answer.

### Reconstruction of the Pavilion and the Scenery

The landscape of The Club at NINE BRIDGES located at the foot of Mt. Hallasan flawlessly with the elegant topography of Mt. Hallasan and the beautiful lines of the golf course. The NINE BRIDGES Pergola takes on the role of providing this scenery by embracing this

최소화시킴으로써 해결될 수 있기 때문이다. 흥미로운 것은 고목에서 시작된 장소성의 재구축을 위한 프로젝트의 논리가 투명한 유리의 물질성과 이를 위한 최소화된 구축 논리를 거쳐 설비와 구조의 실체적인 문제로 귀결된다는 점이다. 이러한 점에서 나인브릿지 파고라는 건축의 지극히 본질적인 장소성의 질문에서 시작되었으나 이를 구현하기 위한 기술적인 진화를 거듭한 프로젝트이다.

### 정자와 풍광의 재구축

한라산 자락에 있는 클럽나인브릿지의 전경은 한라산의 수려한 지세와 골프코스가 지닌 미려한 선들과 절묘하게 조화를 이룬다. 나인브릿지 파고라는 이러한 나인브릿지의 아름다운 전경을 담아내는 데 필요한 풍광 장치의 역할을 수행한다. 즉 유리로 구축된 매스는 고목을 감싸 안으며 새로운 지형을 형성해냄과 동시에 그 형태 자체로서 클럽 나인브릿지의 미려한 풍광을 드러낸다. 투명한 나인브릿지 파고라는 일체화된 건축과 자연의 외부이자 내부이며 전경을 끌어안으며 자신의 존재를 자연의 하나로 치환하는 매개체인 것이다.

이런 의미에서 나인브릿지 파고라는 일종의 건축과 자연을 연계하는 현대적으로 해석된 정자와 같다. 한국의 전통 정원인 소쇄원은 자연의 변화가 입체적으로 펼쳐지는 다양한 공간적, 시간적 풍광들을 즐기는 공간이다. 나인브릿지 파고라 또한 수평적으로 펼쳐지는 제주의 풍광을 입체적으로 담아낸다. 전통 정자의 구축이 보는 소점에 따라 자연적 풍광을 프레임으로 담아낸다면 나인브릿지 파고라 또한 한라산이 지닌 수려한 경관을 투명한 공간적 볼륨으로 받아들인다. 즉 가깝게는 고목을, 멀게는 한라산과 이어지는 골프장의 미려한 전경을 현대적 풍광 장치로서 담아내는 것이다. 이것은 나인브릿지 파고라가 단순히 공간의 기능적 재편뿐만 아니라 3차원적 시간성으로 풍광을 느낄 수 있게 해주는 자연에 감응하는 건축임을 의미한다.

beautiful landscape of the NINE BRIDGES. As the glass mass wraps around the old tree and forms a new terrain, it simultaneously reveals the beautiful scenery of The Club at NINE BRIDGES by itself. Being both a part of the unified structure as well as a part of the scenery, the transparent the NINE BRIDGES Pergola acts as an intermediary by integrating itself with nature.

In this sense, the NINE BRIDGES Pergola is like a modern interpretation of a traditional pavilion that connects architecture with nature. The Korean traditional garden, Soswaewon Gardens, is a place where one can enjoy various spatial and chronological topographies throughout a natural order that unfolds three-dimensionally. Likewise, the NINE BRIDGES Pergola deals with the scenery of Jeju that expands horizontally in a similar manner. Just as a traditional pavilion captures the natural scenery by frames according to the vanishing point, the NINE BRIDGES Pergola receives the beautiful landscape of Mt. Hallasan into its transparent spatial volume. In other words, as a modern scenic viewpoint, it embraces not just the old tree in its proximity but also the beautiful landscapes of Mt. Hallasan and the neighbouring golf course. This means that the NINE BRIDGES Pergola is a structure that can respond not only as a functional reorganisation of space but also as a structure that resonates with the the scenery in terms of three-dimensional time.

### Location as an Extension of Nature

/

If the comfort of the old tree functioned as a basic principle for the arrangement of the NINE BRIDGES Pergola, the design motif that composes its form began with the semantic expansion of the old tree. Through the idea of its locationality, founded on the old tree, the NINE BRIDGES Pergola expands its spatial meaning, and this is then realised through the recomposition of this modern understanding of the old tree within an architectural vocabulary. This suggests a new approach towards space that differs from the traditional

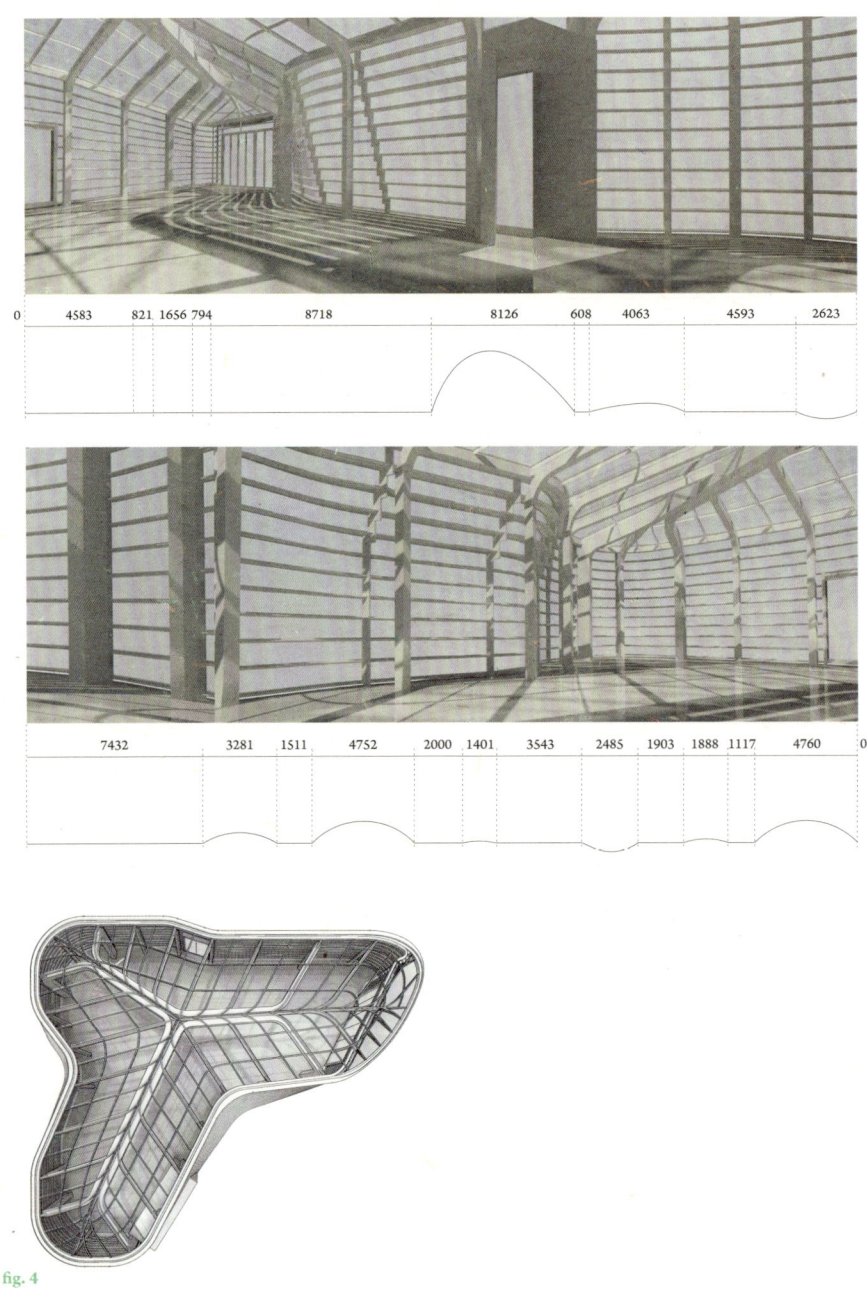

fig. 4

fig. 4 나인브릿지 파고라가 단순히 공간의 기능적 재현뿐만 아니라 삼차원적 시간성으로 풍광을 느낄 수 있는 자연에 감응하는 건축임을 의미한다.

The NINE BRIDGES Pergola is a structure that can respond not only as a functional reorganisation of space but also as a structure that resonates with the the scenery in terms of three-dimensional time.

concept of structure and installation that constitutes space. In other words, if standard architecture divides space in terms of its respective functions by designating structure and installation accordingly, space in nature refers to the unified form itself in which structure and installation are integrated as a whole. To realise this, the architectural volume needs to be able to cater to such a space, and also needs to be able to deal efficiently with the temperamental Jeju climate. For example, Louis Kahn categorised the compositional elements of architectural space into Served Space and Servant Space. In a sense, he divided and analysed the two spaces that comprise architecture—the main space and its subsidiary elements— into what acts as the main element of spatial composition and the supplementary element that supports it. What the NINE BRIDGES Pergola aims to achieve is a space as natural itself – that is, a space where it is not spatially divided by such architectural elements, but can remain undivided as an integrated union between Served Space and Servant Space. This suggests that the logic of the form itself needs to be put in touch with the structural logic, and that the structure must be seamlessly integrated with the flow of the installation. The NINE BRIDGES Pergola depicts architecture as an organism that abides by the logic of nature.

Architecture tends to take inspiration from nature, but in the end, there is always a difference between architecture and nature in terms of its structural logic. The NINE BRIDGES Pergola is a project that not only borrows from nature, but one that strives to resemble it. Through its transparent skin, the NINE BRIDGES Pergola camouflages itself in nature, and simultaneously protects the inner space from the outside. Like the skin of a living organism, glass delivers light and energy to that which lies within, while also dividing the inside from the outside. Moreover, through a ventilation system, fresh air is continuously circulated, and the temperature is moderated through conditioned air during the summer and winter seasons. It has a structure that resembles a living organism which sustains itself in nature. Taking nature as its motif, the NINE BRIDGES Pergola is a project that reflects the form carried within nature into the architectural space.

### 확장된 자연으로서의 공간

고목의 안락함이 나인브릿지 파고라 배치의 근본적인 원칙이었다면 형태를 구성하는 디자인 모티프는 고목의 의미를 확장하는 것에서 시작되었다. 나인브릿지 파고라는 고목과의 장소성을 통하여 공간적 의미를 확장하며 이는 곧 현대적 의미의 나무를 건축적 어휘로 재구성함으로써 구현된다. 이는 전통적으로 공간을 구성하는 구조와 설비의 개념이 아닌 새로운 관점으로 공간에 접근하는 것을 의미한다. 즉 기존 건축이 구조와 설비가 분할되어 공간을 기능적으로 영역화한다면 자연 그 자체로서의 공간은 구조와 설비가 통합된 일체화된 형태 자체를 의미한다. 이를 구현하기 위해서는 건축 볼륨 그 자체가 공간을 마감 지을 수 있어야 하며 제주의 변화무쌍한 기후에 기능적으로 대응할 수 있는 기능적 성능을 만족시켜야 한다.

예컨대 루이스 칸은 건축적 공간 구성 요소를 주 공간Served Space과 부 공간Servant Space으로 구분했다. 즉 건축을 구성하는 주 공간과 보조 공간을 일종의 공간 구성의 메인 요소와 이를 뒷받침하는 보조 요소로 나누어 해석한 것이다. 나인브릿지 파고라에서 구현하고자 하는 것은 이처럼 건축적 요소에 의해서 공간적으로 구분되는 것이 아닌 주 공간과 부 공간이 통합되어 그 경계가 없어진 자연적 요소 그 자체로서의 공간을 의미한다. 이는 형태 그 자체의 논리가 구조적인 논리와 맞닿아 있어야만 하고 구조가 바로 설비의 흐름과 일체화되어야 함을 의미한다. 궁극적으로 나인브릿지 파고라는 자연의 논리로 접근한 하나의 유기체로서의 건축을 의미하는 것이다.

건축은 자연 속에서 많은 영감을 받지만 궁극적으로 건축의 구축 논리는 자연과 건축이 다름을 보여준다. 나인브릿지 파고라는 이러한 점에서 자연에서 영감을 받고 이를 닮고자 시도한 프로젝트이다. 즉 이 프로젝트는 투명한 스킨을 통하여 자연 속에 자신을 위장하고 동시에 외부로부터 내부 공간을 보호한다. 유리는 일종의 생명체의 스킨처럼 빛과 에너지를 전달하며 내외부를 경계 짓는다. 또한 평상시에는 환기 시스템을 통하여 신선한 공기를 순환하고 여름과 겨울철에는 냉난방된 공기로 실온을 유지한다. 이는 마치 자연의 생명체가 생명을 유지하는 방식과 동일한 구조를 지닌다.

## A Double Duct System that Unifies Structure and Installation

The main structural form that cuts across six strands has been divided in three structural directions. This triply divided shape is then structurally divided into six beams, and they deliver the weight of the upper body to six positions. This reveals the formation of a structural form that works as one object, without any division apparent between beams and columns. To realise this natural space, where Served Space and Servant Space have been fully integrated, a double duct system by which structure and installation are unified was developed for this project.

Within the double duct system, the inner duct performs the air circulation and conditioning while the outer duct functions as the structural component. The 12mm thick metal plate is welded to formulate a unified structural frame. The main structural frame that forms an important axis of the interior spacem as a compound curved plate of six strands, fulfills the role as the main duct to account for to the total weight and flow of installation. As a whole, in order to safely control the overall organic shape designed with the compound curved plate, the six strands of the main structural form and 19 pieces of the sub-structural form have been employed.

Moreover, the size of the six main ducts was decided upon according to various complex factors. Considering the regional climate of Jeju, known for its strong winds, a safety factor had to be applied in the cross-section of the structural form, and the air volume in the ventilation and air conditioning system also became a major issue to ponder. When measured under normal air volume conditions, the thickness of the main duct would have become too thick, and this would have undermined the sense of proportion in the interior. To find the appropriate proportion between such practical demands and formal aesthetics, a great deal of time was spent on the architectural design stages of this project. It was important to resolve and find the appropriate solution to the conflict between the required sense of spatiality via

이처럼 나인브릿지 파고라는 자연을 모티프 삼아 자연이 지닌 형식을 건축 공간 속에 반영한 프로젝트이다.

### 구조와 설비가 일체화된 이중덕트 시스템

6가닥으로 가로지르는 메인 구조체는 세 방향의 구조적 흐름으로 분할된다. 3개로 분할된 형태는 구조적으로 6개의 보로 나뉘어 6개의 지점에서 상부의 하중을 전달한다. 이는 보와 기둥의 구분 없이 하나의 오브젝트 자체로서 구조체를 형성하는 것을 의미한다. 이러한 주 공간과 부 공간이 결합한 자연, 그 자체로서의 공간을 구현하기 위해 이번 프로젝트에서는 구조와 설비가 일체화된 이중 덕트 시스템을 고안했다.

이중 관로 중 내부 덕트는 환기 및 공조를 위한 것이며 외부 덕트는 구조체를 구성하는 덕트 역할을 수행한다. 12mm두께의 철판은 용접으로 하나의 일체화된 구조 프레임을 형성한다. 6가닥의 이중 곡면으로 내부 공간의 주요한 축을 형성하는 메인 구조 프레임은 전체적인 하중 및 설비의 흐름을 유도하는 메인 관로의 역할을 수행한다. 전체적으로 이중 곡면으로 처리된 유기적 형태를 안정적으로 제어하기 위하여 6가닥의 메인 구조체와 19개의 서브 구조체가 사용되었다.

또한 6개의 메인 덕트는 여러 가지 복합적인 요소를 통해 크기가 결정되었다. 바람이 거센 제주의 지역적 특성을 고려하여 구조체의 단면에 안전율을 적용해야 했다. 더군다나 환기 시스템 및 에어 컨디셔닝을 위한 공기의 풍량 또한 주요한 이슈가 되었다. 기존의 일반적인 풍량으로 덕트 관경을 해석할 경우 메인 덕트의 두께가 지나치게 두꺼워져 내부 공간의 비례감이 깨질 수 있었다. 이번 프로젝트에서는 이러한 기능적인 요구와 형태미 사이의 적절한 비율을 찾기 위해 건축설계 프로세스에서 많은 시간을 할애했다. 즉 내외부 공간에서 요구하는 투명성으로서의 공간감과 구조적, 설비적 요구에 의한 덕트의 관경 사이의 충돌을 조정하고 최적화된 대안을 찾는 것이

fig. 5 구조와 설비가 일체화된 이중 덕트 시스템을 고안했다.
A double duct system by which structure and installation are unified was developed for this project.

fig. 6 내부 덕트에서 흐르는 차갑고 더운 공기와는 달리 내부공간은 일정한 온도로 유지되기 때문에 온도차에 의한 결로를 해결하기 위하여 고밀도 단열재를 2개의 덕트 사이에 채웠다.
Unlike the cold or warm air that flows within the inner duct, a temperature difference occurs between these two spaces.

■ Supply Air (SA)   ■ Return Air (RA)   ■ Mechanical Room

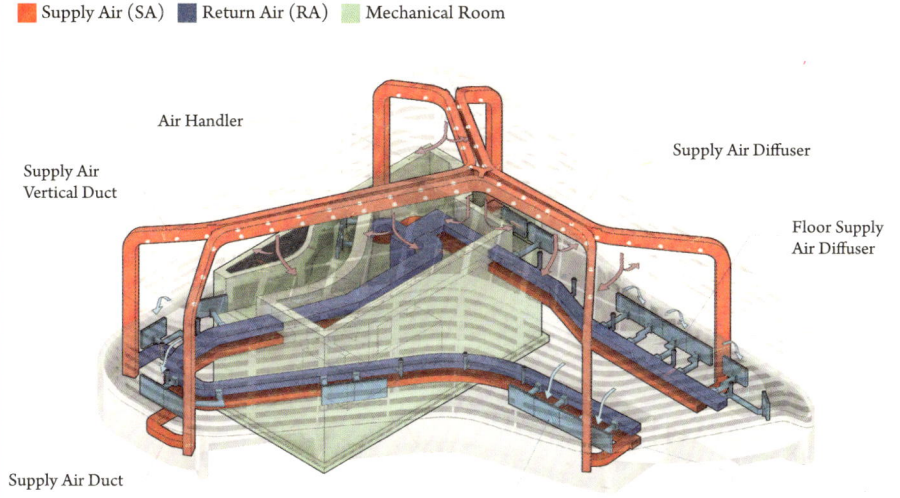

Air Handler
Supply Air Vertical Duct
Supply Air Diffuser
Floor Supply Air Diffuser
Supply Air Duct
Flexible Duct
Exhaust Air B-Line Diffuser

fig. 5

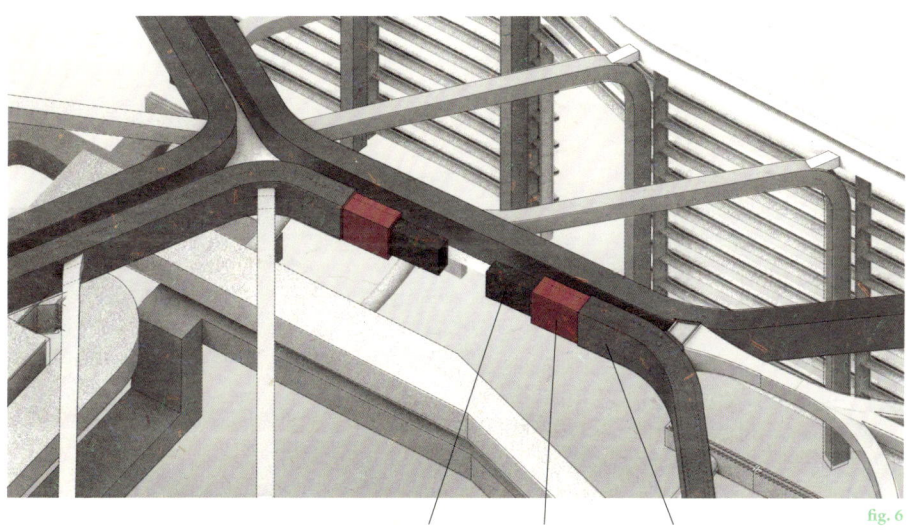

Air Conditioning Duct 400X200    Insulation    Main structure 520X320X12t

fig. 6

중요했다. 또한 여름에는 38°C 겨울에는 -10°C까지 변화하는 제주의 기후에 대응해 내부 공간에 적절한 온도를 유지하기 위해서는 에어 컨디셔닝 상에 발생하는 결로가 예상되었다. 내부 덕트에서 흐르는 차갑고 더운 공기와는 달리 내부 공간은 일정한 온도로 유지되기 때문에 이 두 공간의 온도 차가 발생한다. 이러한 온도 차에 의한 결로를 해결하기 위하여 고밀도 단열재를 2개의 덕트 사이에 채웠다. 또한 설비적으로 사계절의 공간 활용을 위하여 일상적 환기 시스템은 물론 일교차가 큰 제주에서 냉난방이 자체적으로 해결되어야만 했다. 6개의 메인 구조체에 48개의 덕트 배관을 위한 환기 덕트가 설치되었고 이를 통하여 외기에 대응하여 실내의 온도를 일정하게 유지할 수 있었다. 특히 덕트의 풍량만을 고려할 경우 덕트 배관의 크기가 지나치게 커져 구조체의 미려함이 떨어질 수 있어 이를 제어하기 위하여 구조체의 크기와 덕트 배관의 개수 및 공조의 속도를 여러 차례 조율하면서 최적화된 결론을 낼 수 있었다.

### 유리를 이용한 새로운 실험

이러한 구조체 위에 160여 개의 비정형 반강화 복층 유리와 측면의 280여 개의 곡면 유리가 사용되었다. 특히 강화된 단열 기준과 제주도가 지닌 강한 바람과 외기의 변화무쌍한 온도 변화에 대응하기 위하여 좀 더 진화된 타입의 유리를 사용해야 했다. 또한 비용과 기술적인 문제로 인하여 전체 440여 장의 각기 다른 크기의 유리를 중국 공장에서 제작하여 현장에서 제작된 구체에 맞게 조립해야 했기에 작업의 난이도가 높았다. 즉 서울 근교에 있는 공장에서 제작된 내부 구조체는 우선 80여 개의 조각으로 나누어 가조립한 후 다시 제주도로 가져와 재조립한다. 이와 동시에 140여 개의 다른 곡률값을 가진 반강화 이중 곡면 복층 유리는 중국 공장에서 제작하여 한국에서 최종적으로 조립한다.

    용접과 운반을 위해 각각 나누어진 구조체와 유리 개체들은 정확한 데이터 값에 의해서 제작되어야 했고 이를 다시 3D 제작 값과 최종적으로 스캔한 값이 정확히 일치해야만 했다. 자연 그 자체로서의 속성을 지닌 나인브릿지 파고라는

transparency within the interior space and the required duct diameter length in terms of structure and installation. Considering the extremes of the Jeju climate, which reaches 38°C during the summer and -10°C during the winter, the condensation of air was expected to occur during the air conditioning process. Unlike the cold or warm air that flows within the inner duct, because the interior space is maintained at a regular temperature, a temperature difference occurs between these two spaces. To resolve the problem of condensation that would occur due to this temperature difference, a high-density insulation was introduced between the two ducts. In terms of installation, for use of the space throughout the seasons, considering the changeable daily temperature range of Jeju, the air conditioning had to be taken care of with utmost efficiency by the air circulation system itself. 48 circulation ducts for duct piping were installed along the six main structural forms, and through this, the indoor temperature could be maintained at a regular range. The size of the duct pipes could have become too big when considering the air volume of the duct, which would have compromised the structural form aesthetics; but an optimised resolution was reached by mediating the size of the structural form, the number of duct pipes, and the speed of the air conditioning.

### A New Experiment with Glass

Throughout this structure, 160 pieces of atypical semi-tempered multi-panel glass were used and 280 pieces of curved glass were used at the back of the structure. In order to respond to the enforced insulation standards, as well as to the strong winds and the temperamental climate of Jeju, a more progressive type of glass was required. Furthermore, due to the cost and technical problems, the 440 pieces of glass of varying sizes had to be manufactured at a factory in China, and there was also the difficulty of having to assemble them to fit the structural form. The inner structural form, which was manufactured at a factory nearby in

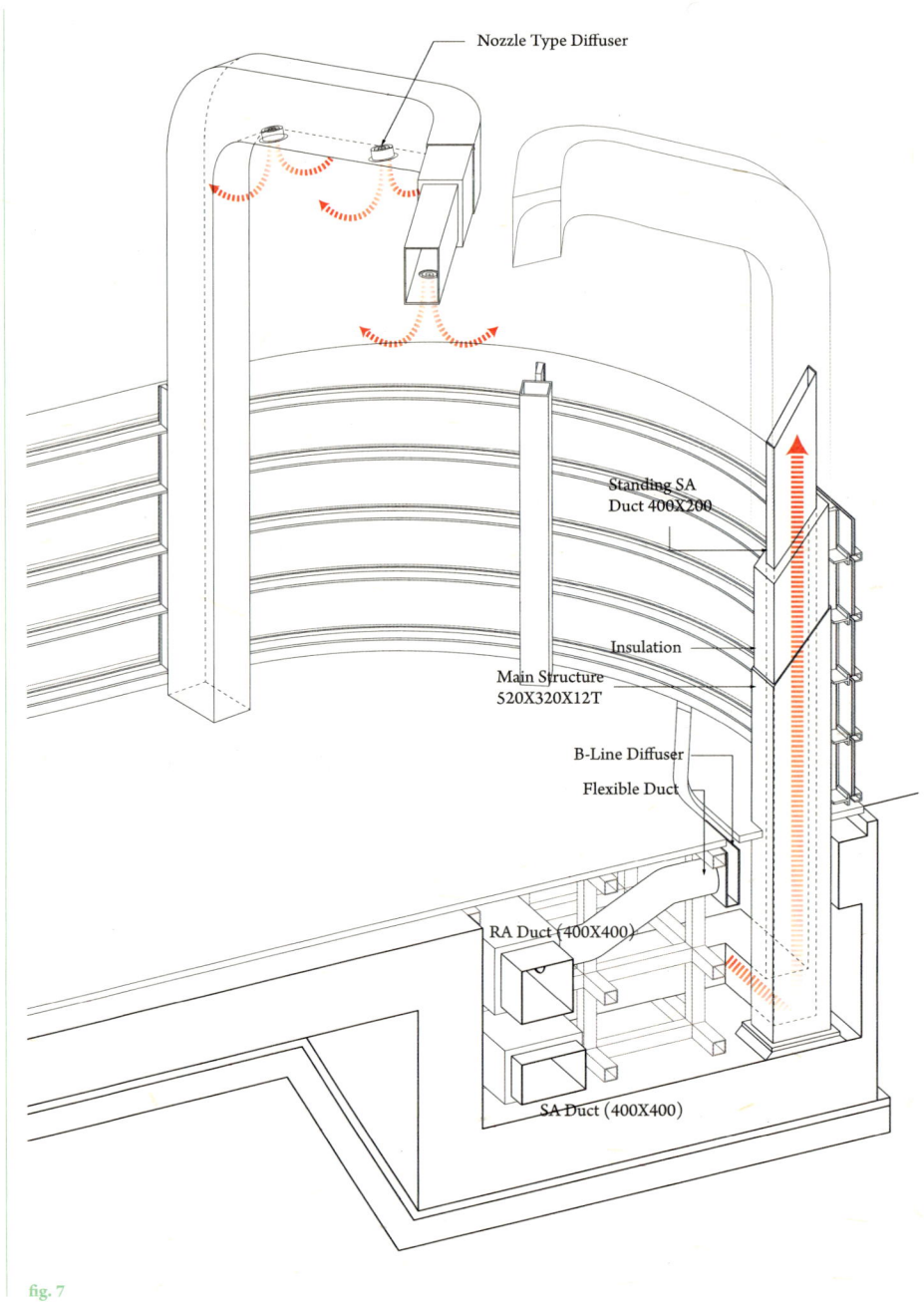

fig. 7

fig. 7 내외부 공간에서 요구하는 투명성으로서의 공간감과 구조적, 설비적 요구에 의한 덕트의 관경 사이의 충돌을 조정하고 최적화된 대안을 찾는 것이 중요했다.

It was important to resolve and find the appropriate solution to the conflict between the required sense of spatiality via transparency within the interior space and the required duct diameter length in terms of structure and installation.

Seoul, had to be dismantled into about 80 pieces, and after being tentatively assembled, they were brought to Jeju and reassembled there. Simultaneously, the semi-tempered double composite multi-panel glass, with approximately 140 different curvature values, was manufactured in a Chinese factory and was finally assembled in Korea.

The structural form and the glass pieces that were divided for welding and transportation had to be manufactured according to a precise data value, and this presented difficulties as the values had to coincide exactly with the 3D production value and the final scan value. Borrowing its character from nature itself, the NINE BRIDGES Pergola is an atypically-shaped architectural structure, that is based on a three-dimensional curvature. Despite the fact that its materials were produced in two different countries, the project demanded an extremely precise working method in which there was no room for error – any errors during the assembly phase were not to exceed 10mm.

For this exacting work, the project was to proceed under a close relationship between the Architectural office and the construction work done in the field, with the 3D based programme as mutual reference point. The design process, based on BIM and on-site shop drawing, demanded a unified file sharing system for the sake of data expansion and congruence. Furthermore, without knowledge of field production and technical expertise, the work during the construction phase would have been very difficult. The fact that all building materials were produced at factories and then brought on site to be assembled offered the possibility that all these materials could be dismantled and reassembled again somewhere else other than Jeju. Looking ahead, this also suggests the possibility for a new mode of construction that goes beyond the traditional methods of construction.

3차원 밴딩을 기반으로 한 비정형 건축이다. 이는 모든 부재가 한국과 중국 두 곳에서 제작되었음에도 현장에서 다시 재조립되었을 때 오차가 10mm를 넘어서는 안되는 정교한 작업을 요구하는 것이다.

이러한 정교한 작업을 위하여 3D 기반의 프로그램을 바탕으로 설계사무소와 현장 사이의 긴밀한 관계 속에서 작업이 진행되었다. BIM 기반의 설계 프로세스와 현장에서 제작된 숍드로잉(현장도면)은 데이터의 확장과 연속성이라는 측면에서 일원화된 파일 공유 시스템을 요구했다. 무엇보다도 기술적인 숙련도와 현장제작의 노하우가 없었다면 현장에서 공사 기간 내에 작업하기 어려웠을 것이다. 즉 프로젝트에 소요된 모든 구조물을 공장에서 제작하여 현장에서 조립했다는 것은 모든 구조물을 다시 분해하여 제주가 아닌 다른 곳에도 정확히 재현할 수 있다는 뜻이다. 이는 좀 더 미래적 의미에서 전통적인 방식의 공사 시스템을 극복하고 새로운 방식의 공사 과정의 가능성을 의미하기도 한다.

### 구축 체계의 새로운 질서의 가능성

역사적으로 건축가들은 자연과 닮은 건축을 꿈꿔왔다. 모방의 대상으로서 자연은 언제나 잡힐 수 없는 근본적으로 다른 경외의 대상이었다. 그것은 건축이 가진 근본적인 한계이자 창조주와 인간의 차이를 드러내는 접점이기도 했다. 이러한 의미에서 나인브릿지 파고라는 단순히 형태적으로 재구축된 유리 파빌리온을 의미하지 않는다. 이것은 새로운 형태 실험을 넘어 구조적, 설비적으로 일체화된 의미의 자연화된 공간을 의미한다. 전통적인 의미의 벽과 슬라브, 천장은 하나의 스킨으로 일원화된다. 벽은 곧 투과성 있는 천장이자 천장은 곧 벽의 연속된 개체이다. 유리로 일원화된 스킨은 그 자체로서 내외부 공간을 분할하며 시각적으로 외부 공간을 내부화하는 매개체 역할을 한다. 그것은 창이라는 이차원적 입면의 오프닝을 통해서 자연을 인식하는 감성을 넘어 3차원적 볼륨 속에서 보다 입체적으로 자연의 감성을 받아들일 수 있음을 의미한다.

## Possibility for a New Order in Construction

Historically, architects have dreamt of an architecture that resembles nature. As an ideal, which a building may strive to attain, nature has ever been an object of awe that could never be fully grasped. That is the fundamental limit of architecture, and also the point of difference between God and man. In this respect, the NINE BRIDGES Pergola is not merely a glass pavilion that is reconstructed as an illuminating form. Instead, it points towards a naturalised space – a space that goes beyond a mere experiment in form to arrive at a unification of structure and installation. The traditional wall, slab, and ceiling are integrated into the skin. The wall is a ceiling with a penetrable façade, and the ceiling is a continuous aspect of this wall. The skin has integrated with the glass and segregates the outer space, functioning as a medium that interiorizes the inner space a virtual way. This suggests that the natural aspects of the form can be experienced more realistically in a three-dimensional volume that goes beyond a mere perception of nature, made accessible from a 2D opening in the transparent plane like a window.

In this way, the project can be understood to have made several new attempts in terms of developing a spatial positioning logic, formal composition, and a degree of technical completion. While the NINE BRIDGES Pergola began by questioning the relationship between nature and architecture, it seems that the method of its realisation was made possible by the fulfillment of technical developments. Such architectural accomplishments have presented the industrialisation of specialised technical aspects in an atypical architecture, aspects that are not generally adopted by the market. Moreover, this process of development contains the possibility for an industrial system that will be able to ingest the new form and to refine it with continued skill. In this way, an architect evolves along with the architectural industry, and produces not only new meanings of space but also industrial advancements. In other words, while architecture creates new meanings for the land with humanity as its

이러한 관점에서 이번 프로젝트는 공간의 배치 논리와 형태 구성, 그리고 기술적 완성도까지 많은 부분이 새롭게 시도되었다고 평가된다. 나인브릿지 파고라는 자연과 건축의 관계성에 대한 질문으로부터 시작되었지만 그 구현의 방식은 진화된 기술적 성취로 보는 것이다. 이러한 건축적 성취는 일반적으로 양산되지 않은 비정형 건축의 특화된 기술이 그 시장 속에서 산업화함을 의미한다. 또한 이러한 진화의 과정은 일정한 사회가 지닌 산업 시스템이 새로운 양식을 소화하고 이를 능숙하게 재현할 수 있는 가능성을 내포한다. 이처럼 건축가와 건축산업은 서로 맞물려 진화하고 새로운 공간의 의미뿐 아니라 산업적 확장성을 생성해낸다. 즉 건축은 인문학적 토대를 바탕으로 대지에 의미를 만들어내지만 궁극적으로는 기술적 진화를 통한 산업화의 결정체인 것이다. 나는 이러한 맥락에서 나인브릿지 파고라에 또 다른 의미를 부여하고 싶다.

나는 이와 더불어 유리 건축이 가진 한계와 이를 극복하는 과정에서 찾아낸 몇 가지의 가능성에 주목하고자 한다. 즉 온실의 기능적 요구에서 시작된 유리 건축은 끊임없이 진화하여 현대건축의 근간을 이루고 있다. 이러한 투명성을 근간으로 하는 미학적인 이유와 더불어 주목해야 할 점은 조립과 해체를 통한 유리 건축의 재구축성이다. 3차원적 이중 곡면을 지닌 유리의 조각들은 그 자체로 빛의 투과성과 단열성을 확보할 수 있는 주요한 건축 재료이다. 즉 기존의 콘크리트나 철골 건축에서 요구하는 구조재 이외의 단열재 및 마감재 없이 그 자체로 유닛이 마감되는 독특한 구축 형식을 지니는 것이다. 이것은 달리 말하면 조립과 해체의 방식에 따라 다른 재료가 지닌 고정적 구축성과는 다른 공간적 이동의 유연함을 의미한다. 또한 이와 연계된 설비와 구조 시스템은 유리의 시공성과 더불어 구축적 용이함에서 좀 더 다른 방식으로 진화할 가능성을 내포하고 있다. 즉 일체화된 구조와 설비 시스템은 기존의 건축 공간에서 불가피하게 분할하여 고려해야만 했던 덕트 개념을 새롭게 해석하여 공간 확장의 다른 가능성을 낳는 것이다.

자연에 대한 존중으로부터 시작된 나인브릿지 파고라는 공간 구성의 새로운 논리를 제시한다. 자본화된 현대 도시의 생성 논리와는 반대로 건축과 자연의 관계에

grounding, ultimately, however, it is also the fruit of industrialisation garnered from such technical evolutions. I would like to add this aspect as emerging from the context of the NINE BRIDGES Pergola.

Along with this, I would like to focus on the new possibilities that have emerged during the process of exploring and overcoming the limits of glass architecture. Glass architecture, which began with the practical request of a greenhouse, has continuously evolved and has become a part of the foundation of modern architecture. The point that we should focus on, along with the aesthetic outcomes of the glass's transparent quality, is the reconstructability of glass architecture, via assembly and dismantlement. The glass pieces, with 3D double composite curved surfaces, are an important architectural factor that can secure both the penetrability and insulation of light. It takes a unique structural form, where it as a unit is made complete without the need for insulation or finishing materials other than the building materials used in regular concrete or steel frame architecture. In other words, this implies a flexibility of spatial movement enhanced by its method of assembly and dismantlement, which differs from the fixed structural properties embodied by other materials. Also, along with the constructability of glass, the related installation and construction system contains within it the possibility to evolve in a different way in terms of a construction facility. It is only when a unified structure and installation system is divided and considered within a standard architectural space that the concept of the duct can be newly interpreted and prompt a new means of spatial expansion.

The NINE BRIDGES Pergola, which began with a sense of respect for nature, proposes a new architectural approach to spatial composition. In contrast to the commercial logic of a contemporary urban city, it opens up a new perspective on the relationship between architecture and nature. As if mocking the logic of the contemporary urban city, which seeks to judge everything under rationality and economic value, it proposes a new generative grammar and thereby inspires questions about the essence and meaning of architecture. The

새로운 관점을 열어젖힌다. 그것은 합리성과 경제성으로 모든 것을 판단해야만 하는 현대 도시의 논리를 비웃기라도 하듯이 새로운 생성 문법을 제시하며 건축의 본질과 의미에 새로운 질문을 던지는 것이다. 내 기억 속의 나무는 자연의 일부분이 아닌 한 마을의 집단적 기억을 공유하게 하는 구심점이었다. 이곳에 위치한 나무 또한 클럽나인브릿지를 방문한 사람들을 모으고 통합하며 한 공간의 일원이 되었음을 상징화한다. 클럽나인브릿지의 유려한 풍광과 제주의 자연은 고목과 파고라를 통해 재구성되며 건축과 자연의 새로운 관계 맺음으로 기억될 것이다.

tree of my memories was not a part of nature but a central pivot that allowed for the sharing of the town's collective memory. This aging tree also brings together and unifies the visitors to The Club at NINE BRIDGES and signals that they are now members of the same space. The beautiful scenery of The Club at NINE BRIDGES and the nature of Jeju are recomposed through the tree and the Pergola, and they will be remembered in the new relationship struck between architecture and nature.

Drawn by JOHO Architecture

피포 쵸라
로마국립21세기미술관
건축 선임 큐레이터

# 호흡하는 파빌리온

## The Pavilion that Breathes

Pippo Ciorra
Senior Curator,
MAXXI Architecture

클럽나인브릿지 파고라는 형태와 기술의 기로에 선 특별한 프로젝트다.
건축가는 이 프로젝트를 통해 파빌리온 건축의 독특한 패러다임 중 일부를
재정립하려는 것이다. 건물을 구성하는 다양한 선들은 고대의
나무와 경관을 향한 존중의 형태를 그리는 사랑스러운 몸짓이다.

### 제주도, 한국, 건축가 이해하기

「SPACE(공간)」로부터 이정훈 건축가의 나인브릿지 파고라의 비평을 요청을 받았을 때 나는 매우 걱정스러웠다. 지난 몇 년간 한국에 자주 다녀왔음에도 불구하고 제주도에 한 번도 가본 적이 없었고, 나인브릿지 파고라가 건설된 제주도의 역사나 지리도 몰랐기 때문이다. "사소한 것에서 디테일한 것을 찾아야 한다"는 말처럼, 의미를 정확히 살펴보아야 하는 작고 단순한 건물에 대한 비평은 매우 어려운 일이다. 게다가 이 작은 건물이 파빌리온일 경우 대부분은 임시로 행사를 하는데 쓰므로, 나인브릿지 파고라를 비평하는 일은 더욱 도전적인 임무였다. 그래서 나는 좀 더 시간을 갖고 생각해보았다. 그 후 몇 가지 새로운 요소들을 논의했고, 편집자의 조심스러운 주장이 나의 관심을 끌었다. 나의 관심을 사로잡은 요소는 바로 편집자가 반복적으로 언급한 모든 근대 파빌리온의 건축적 아버지 조셉 팩스턴의

수정궁the Crystal Palace이다. 처음에는 조금 달갑지 않았다. 규모, 시기, 맥락, 기술 등 두 개의 매우 다른 요소를 비교하는 오만함(혹은 순진함) 때문이었다. 그러다가 건축가를 만나 이야기를 나누어볼 기회가 생겼고, 그는 나와 프로젝트에 관하여 의견을 나누고 도면들을 보여줬다. 그리고 나는 형태와 기술의 기로에 선 특정한 관점에서 이 프로젝트를 바라보면서, 그가 이 프로젝트를 통해 사실상 파빌리온 건축의 특정한 패러다임 중 일부를 재정립하려 한다는 것을 이해할 수 있었다. 이 글은 '호흡하는 파빌리온'이라는 관점에서 파빌리온 건축으로서 특성과 혁신을 생산하려는 이 프로젝트의 몇 가지 측면들을 논의하려고 한다.

### 파빌리온 건축의 패러다임

첫 번째 주제는 안정성과 유동성 개념이다. 만약 당신이 알바니아에 있는 티라나를 산책한다면, 2013년 런던에서 전시되었던 후지모토 소우의

**fig. 1** 근대 파빌리온들의 건축적 아버지 조셉 팩스턴의 수정궁은 나인브릿지 파고라와는 규모, 시기, 맥락, 기술 등에서 매우 다른 요소를 가지고 있다. Paxton's the Crystal Palace, which is for many the architectural father of all modern pavilions. It has very different elements from the NINE BRIDGES Pergola, such as size, timing, context, and technology.
©Philip Henry Delamotte

**The Pergola of The Club at NINE BRIDGES is a special project that becomes interface between form and technology. The architect aims to redefine the part of the unique paradigm of Pavilion architecture via the project. The various lines that consist the building are lovely gestures that ode to ancient trees and landscape.**

## Understanding Jeju, Korea, Architects

When I received an invitation from *SPACE* to write a critique of Lee JeongHoon's the NINE BRIDGES Pergola in Jeju, I felt very dubious about the prospect. Despite my frequent trips to South Korea in the past few years, I had never been in Jeju and I did not know the history and geography of the place in which the pavilion was under construction at that time. Aside from these concerns, it has to be said that it is always a hard task to write critically about small and simple buildings, where meanings have to be looked for with a very accurate eye, and 'God…' – as it used to be said – '…has to be found in the details'. The mission becomes even more challenging when the small building is a pavilion, which is – in most cases – basically a roof to host temporary activities. So I took some time to think it over. Then a few new elements were introduced into the discussion, and brought to my attention by the *SPACE* editors' gentle insistence. The first detail

fig. 1

to catch my attention was a recurring reference in the editors' words to Paxton's the Crystal Palace, which is for many the architectural father of all modern pavilions. In the beginning I was a bit disturbed. There could be a trace of arrogance (or naivety) in *SPACE*'s comparison of two projects of such different scale, date, context, technology, and ambition. Then I had the chance to meet the architect and to talk to him. He discussed the project with me and showed me the drawings. I understood that from a specific point of view, suspended between form and technology, the project of Lee JeongHoon is in effect trying

fig. 2

fig. 2 (위) 2013년 후지모토 소우는 켄싱턴 가든에 서펜타인 파빌리온을 디자인했다.
(아래) 후지모토 소우의 서펜타인 파빌리온은 수년간 티라나에 대여되었다.
(top) Sou Fujimoto designed Serpentine Pavilion for gallery in Kensington Gardens. ©Iwan Baan
(bottom) Fujimoto's work has been leased to the city of Tirana for a number of years.

to redefine some of the specific paradigms of pavilion architecture. So the task of this text will be to look at *The Pavilion That Breathes* from this point of view, in an attempt to discuss aspects of a project that tries to prompt innovation and devise new qualities to the ideas that inform pavilion architecture.

**Paradigms of Pavilion Architecture**

The first topic I would like to address within this frame is the idea of stability vs mobility. If any of you have taken a walk across the city centre of Tirana, in Albania, he or she may be surprised and displaced by the view of something that looks very much like a copy of Sou Fujimoto's 2013 Serpentine Pavilion in London. Coming closer to take a serious look at the object, we realise, to some surprise, that it is not a copy. It is the actual installation designed for the lawn in front of the well-known gallery in Kensington Gardens. After the summer of 2013, Fujimoto's work has been leased to the city of Tirana for a number of years and for a space which has very little to do with the original commission. The reference is useful only to demonstrate the degree to which we have gotten used to considering pavilion architecture of recent years as closer to works of art. It is often a museum project—as for the Serpentine, YAP, many Biennales—that has no programme and the only task set is to display the author's architectural manifesto, which corresponds with what happens when an artist is asked to produce a tridimensional installation for an exhibition. As a museum 'item', or installation, the contemporary pavilion embodies and develops the old idea of temporariness and transportability, as in the North-African marriage tents or even in the Crystal Palace itself, which was displaced somewhere else after the Expo (and after it, destroyed by a fire). The brief for the Serpentine Gallery warns designers that their project has to be easily re-displayable at other locations. The 'item' can travel from one exhibition to another, since the pavilion has mostly become an item connected to an exhibition.

The NINE BRIDGES Pergola has very little to do with this. The two main elements cited in the architect's presentation are programme and place. The programme is deeply contextual, and it has to do with the rest of the buildings hosting the activities of the local golf club. The project provides the golf club with an additional open space for parties and events, which the club was missing. It does this by speculating on the form of the existing buildings and their relationship to the landscape, and then displaying a form that is both

서펜타인 파빌리온이 있는 것을 보고 깜짝 놀랄 것이다. 그러나 티라나에 있는 파빌리온에 가까이 다가가 자세히 들여다보면, 뜻밖에도 모조품이 아니라는 사실을 깨닫게 된다. 그것은 켄싱턴 가든에 있는 미술관과 잔디밭을 위해 디자인되었던 바로 그 설치물이다. 2013년 여름 이후, 후지모토의 작품은 켄싱턴 가든과 전혀 관련이 없는 공간인 티라나에 수년간 대여되었다. 이러한 사례는 우리가 어쩌다가 지난 몇 년간 파빌리온 건축을 예술품이라 여기게 되었는지 설명해주는 대목이다. 보통 서펜타인 갤러리 파빌리온, YAP Young Architects Program 혹은 많은 비엔날레들을 위해 지어진 미술관 프로젝트는 작가의 건축적 선언문을 전시할 때 외에는 특별한 프로그램이 없다. 한 작가가 전시회를 위하여 3차원의 설치 미술품을 제작할 때도 마찬가지다. 미술관 작품, 혹은 3차원 설치 미술품으로서의 현대 파빌리온은 북아프리카의 결혼식 텐트나 엑스포 이후 다른 곳으로 전용되었던 수정궁처럼(그리고 나서 불에 타버렸지만), 일시성과 운반성을 구현하면서 발전되었다. 서펜타인 갤러리의 지침서에는 디자이너(건축가)에게 파빌리온이 다른 곳에서도 용이하게 전시될 수 있어야 한다고 명시되어 있다. 이제는 파빌리온이 일반적으로 전시의 작품이 되어버렸기 때문에, 작품은 응당 다른 전시로 이동이 가능해야 한다.

하지만 나인브릿지 파고라는 이런 맥락과 관련이 거의 없다. 건축가의 설명에서 인용된 두 가지 주요 요소는 프로그램과 장소다. 여기서 프로그램은 맥락과 깊은 관련이 있으며 현지 골프 클럽시설과 연결된 나머지 건물들과도 관련이 있다. 나인브릿지 파고라는 골프클럽 내에는 없는 파티나 이벤트를 위한 오픈스페이스를 제공한다. 그것은 기존 건물들과 조경의 형태를 고려하여, 부드러우면서 완전한 자유로움을 보여주며 나머지 건물에 기타 시설과 서비스 공간이 의존하는 형태를 보여준다. 건축가에게 이 장소와 경관의 관계는 600년도 더 된 팽나무의 존재로 확인되며, 이는 새로운 파빌리온의 형태와 본질을 만들어내는 과정의 주된 근거가 된다. 건축가가 이 프로젝트에 접목한 또 다른 단서는 고대 그리스 델피의 신탁이라는 점이 흥미롭다. 수정궁이 건축에서 수익과 자유로운 교환을 위해 언제라도 이동 가능한 시장경제의 첫 번째 구체적인 단서라면, 나인브릿지 파고라는 기념비적인 팽나무로 인해 수호신을 추구하는 사람들과 자연에게 '신전' 같은 역할을 한다. 그러나 우리는 건설 과정이 보여주는 기술과 '현대적인' 개념에서 어떠한 향수나 회귀적인 건축 양식을 찾아내지는 못했다.

건축가의 아이디어를 충분히 이해하기 위해서는 그가 선택한 기술, 재료 그리고 시공 방식이 이 프로젝트 내에서 불안정성, 이동성과 어떻게 관계를 맺었는지 알아야 한다. 설계는 장소특수성을 강하게 드러내지만 건축가는 그 대지 깊숙이 내재되어 있는 불안정함과 일시성을 전적으로 없애려 하지 않았다. 프로젝트 설명

soft and completely free, clearly relying on the rest of the complex for facilities and service spaces. For Lee JeongHoon, the relationship of the structure to the place and to the landscape is mostly connected to the presence of the old tree, a hackberry tree more than 600 years old, which becomes the main reason behind generating the shape and essense of the new pavilion. It is interesting that another clue offered to architecture fans by the architect is the reference to the Delphi Shrine of ancient Greece. Where the Crystal Palace remains the first concrete sign of the presence of market economy in architecture, ready to pack up and move to another site for the sake of profit and free exchange, the NINE BRIDGES Pergola in Jeju is a 'temple' for people seeking nature and some form of animist genius loci, generously offered by the monumental huckberry tree. However we won't trace any nostalgia or regressive architectural feeling in the project, thanks to the 'modernity' of its form and, above all, to the technology implied by the construction process.

To achieve a full understanding of JeongHoon's ideas we must also note how the choices in terms of technology, materials, construction helped to ambiguously re-engage with the idea of instability and mobility in the project. The design is firmly site-specific but the architect does not want to completely erase the idea of instability and temporariness, so deeply embedded in the architecture of this part of the world. As we read and understand from the documentation, a large part of the structure was built far from the site. The glass panels were produced in China and the loading structure was produced and assembled near Seoul. Basically the NINE BRIDGES Pergola, as many of the nomadic structures to which we have been referring, were built in a workshop, then disassembled and re-assembled in Jeju. In the end this ambiguity between stability and mobility becomes one of the more charming aspects of the project, amplifying the structure's depth and sophistication.

The second aspect I would like to address has to do with the beneficial moves made by the architect when transforming the technical and structural elements of the project into architectural qualities. Looking at the plan it becomes very clear that the space is completely empty, as with a real pavilion. No restrooms, no kitchen, no storage space. Everything that could be eliminated or reallocated elsewhere has been removed from the space. The only presences Lee JeongHoon cannot free from the space from are structure and the climate control, and these aspects reveal his decision to merge the two systems to inform the determining identity of the project, together with the surrounding landscape. The loading

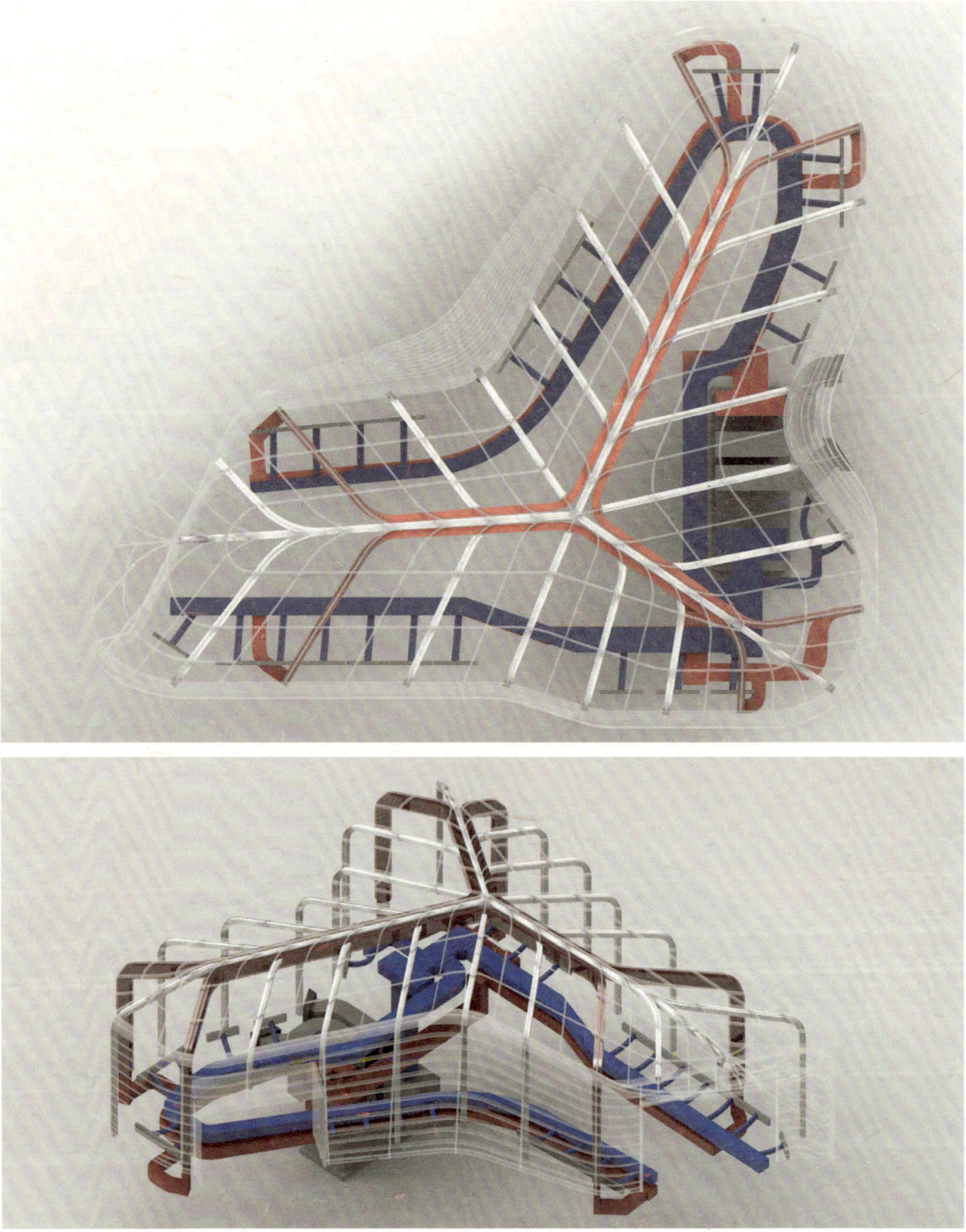

fig. 3

**fig. 3** 나인브릿지 파고라의 구조는 3차원적인 잎을 닮았고, 실내온도 조절을 위하여 철과 유리로 만들어진 큰 잎의 '잎맥들'을 따라 흐른다.

The NINE BRIDGES Pergola's structure clearly resembles that of a three-dimensional leaf, so it is no surprise that we find the fluids of the climate control running through the 'veins' of the large leaf made of steel and glass.
Drawn by JOHO Architecture

structure and climate ducts overlap, intertwine and collaborate in their articulation of the space. The NINE BRIDGES Pergola's structure clearly resembles that of a three-dimensional leaf, so it is no surprise that we find the fluids of the climate control running through the 'veins' of the large leaf made of steel and glass. As viewers, we're fascinated by the way beams and ducts interweave to create a new organic idea of tectonics.

Observing the plan we can observe some of the other interesting aspects to the project. The first is the distance between the pillars and the elevation, growing towards the ground and lending great importance to the elements of the glass façade. The second is the way 'nature' penetrates the building and enters the space that has just been highlighted, between structure and the glass façade. It has two main consequences: it emphasises the main ambition of the pavilion as being at one with nature, and it gives even greater importance to the only 'object' in the space and camouflaged by the structure, the devices that control the climate. This allows not only human guests but also the natural world to inhabit the building without disease.

The third point is obviously the 'formal' one. The first strong impression we receive when we look at the images and the plan are again that the NINE BRIDGES Pergola is a very unusual work of pavilion architecture. The main feature in such projects is generally the idea of the isolated object placed on a green or a more generic site. The 'detached' condition of the pavilion is not only a feature of its aesthetic but is also functional. Mobile and temporary architecture has to be easily disassembled and quickly assembled, and so free space around the structure is essential. In this respect, Lee JeongHoon chooses to challenge the nature of pavilion architecture. His object is designed to remain and it is meant to establish close and even conflicted relationships with the existing complex. It would be reductive to consider it an 'addition'. With his smooth moves, JeongHoon wants to re-discuss (and re-construct) architectural thinking behind the club. His project forces its way between three existing bodies: the two pre-existing buildings and the previous glass addition to the party lounge. There's no pursuit of the individuality of the object or of the easy isolated identity of the project. The architect clearly aims at changing the architecture of the whole complex, re-defining it around the strong and ancestral presence of the hackberry tree.

글에서 알 수 있듯이 구조물의 재료 대부분은 대지에서 멀리 떨어진 곳에서 제작됐다. 유리 패널은 중국에서 생산되었으며 하중 구조물은 서울 인근에서 생산되고 조립되었다. 기본적으로 나인브릿지 파고라는 지금까지 우리가 언급해온 대부분의 파빌리온 구조물과 마찬가지로 공장에서 생산되어 조립 후 해체되었다가 제주도에서 재조립되었다. 결국 이러한 안정성과 유동성 사이의 애매함은 이 프로젝트의 매력적인 부분이 되었고, 그 깊이와 치밀함을 증폭했다.

두 번째 측면은 건축가가 기술적이고 구조적인 요소를 건축적 품질을 높이는 수단으로 사용했다는 점이다. 평면을 살펴보면 내부 공간이 실제 파빌리온처럼 비어 있다는 것을 알 수 있다. 화장실도 없고, 주방도 없으며, 창고도 없다. 철거되거나 다른 곳으로 이동될 수 있는 모든 것들은 일찍이 이 공간에서 없어졌다. 건축가가 유일하게 벗어날 수 없었던 것은 구조와 온도 조절이었다. 그래서 건축가는 주변 경관과 맞물려 이 두 시스템들을 조합하는 것이 이 프로젝트의 주된 정체성을 만드는 것이라 결정한 듯 보인다. 하중 구조물과 실내온도 조절장치는 공간을 연결하기 위하여 포개지고 뒤얽힌다. 나인브릿지 파고라의 구조는 3차원적인 잎을 닮았고, 실내온도 조절을 위하여 철과 유리로 만들어진 큰 잎의 '잎맥들'을 따라 흐르는 형태는 놀라운 일도 아니다. 방문객은 보와 배관이 구축의 새로운 유기적 아이디어를 만들기 위해 섞여 있는 모습에 매료될 것이다.

평면을 살펴보면, 우리는 또 다른 새로운 측면들을 발견할 수 있다. 하나는 유리 파사드를 구성하는 요소에 중요성을 부여하면서 땅을 향해 성장하고 있는(건물을 잎으로 본다면) 기둥과 입면 사이의 거리다. 또 다른 하나는 앞에서 강조한 구조와 유리 파사드 사이의 공간을 관통하는 자연이다. 이 두 가지는 주요한 결과를 가져온다. 자연과 결합한 파빌리온의 주된 포부를 강조하고, 실내에 있는 구조에 위장한 유일한 '대상'에 더 큰 중요성을 강조한다. 그 대상은 온도를 제어하고 사람들이 방문할 수 있도록 허용할 뿐만 아니라 자연이 건물에 서식할 수 있도록 허락한다.

이 프로젝트의 세 번째 관점은 너무도 명백하게 '형식'이다. 프로젝트의 이미지들과 평면을 보았을 때 가장 강하게 받는 인상은 재차 언급하지만 나인브릿지 파고라가 매우 특이한 파빌리온 건축이라는 점이다. 파빌리온 건축의 가장 주된 특징은 일반적으로 녹지나 지면에 배치된다는 점이다. 대부분 파빌리온의 '무심한' 태도는 미학적인 특성일 뿐만 아니라 기능적이다. 유동적이고 일시적인 건축은 빠르게 조립하고 분해해야 하기 때문에 주변에 아무것도 존재하지 않아야 한다. 하지만 나인브릿지 파고라의 건축가는 파빌리온 건축의 본질에 도전했다. 이 프로젝트는 머물러 있기

**fig. 4** 세련된 이중곡선의 유리 파사드는 새로운 퍼포먼스를 제공한다. 건물 안에 있는 사람들의 시선이 나인브릿지 파고라에서 경관으로 주변 건물들로, 방해받지 않고 움직일 수 있다.
The sophisticated double curve glass façade, produced through ambitious construction processes, provides the first performance. It allows our views to move undisturbed from the NINE BRIDGES Pergola to the landscape to the old buildings to the previous extension.
Drawn by JOHO Architecture

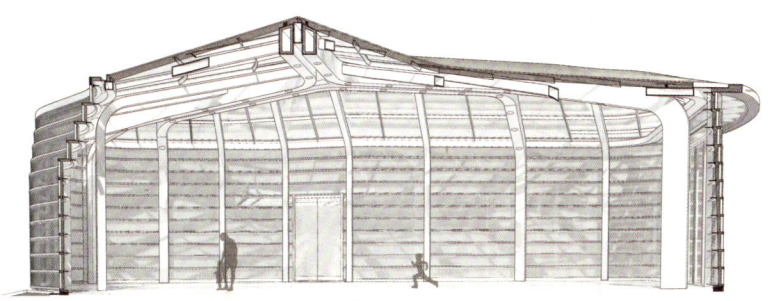

fig. 4

### Glass Box with a Metal Structure in 21C

Such a task is achieved through a series of simple moves. The first is the choice of the informal, leaf-like shape of the pavilion. There's no analogue geometric hierarchy, nor is there an evident digital algorithm guiding the lines. They spring by a lovely gesture of trying to draw a form of respect towards the ancient tree and the landscape. The diagram is not programme-oriented: the trilobal layout clearly puts form before function. The project's main task is to privilege the marriage between architecture (and its guests) and nature rather than filling the space with as many tables as possible. On the other hand, the landscape-oriented choice of the three wings ends up making the space more agreeable, with a lower sense of crowding and a better management of potential noise. The second design move worth mentioning here is the continuous and transparent nature of the walls. Again, Lee JeoungHoon aims at a double and slightly contradictory goal. He wants total transparency, but he also wants the form of his leaf not to disappear completely in the fluidity of the space. The sophisticated 'double curve' glass façade, produced through ambitious construction processes in CJ E&C (CM) and ILJIN Unisco (Steel Structure), provides the first performance. It allows our view to move undisturbed from the NINE BRIDGES Pergola to the landscape to the old buildings to the previous extension. This last feature will be important, especially when parties involve both lounges and visitors will be encouraged to feel like they belong to the same feast. However, the glass façades are marked and limited by a very solid frame of horizontal elements, completing the re-construction of the everted skeleton in our leaf-like building, reinstating individuality to the new pavilion. The third and somehow obvious choice, which completes the overall image of the project, is the choice of an 'everted' structure, which is clearly visible through the façades and the glass roof. This provides the fullest sense of openness and

위해 디자인되었고 기존 건물과 가깝거나 때론 대립되는 관계들을 고려했다. 이것을 '증축'이라고 한다면 환원주의적일 것이다. 건축가는 매끄러운 움직임과 함께 클럽 내 모든 건축을 재론(재건축)하고 싶어 했다. 그의 프로젝트는 3개의 기존 건물들(이미 존재하고 있던 2개의 건물들과 파티 라운지에 추가된 유리 구조물) 사이에서 자체적 방식을 강요한다. 하지만 객체의 특성이나 쉽게 분리된 프로젝트의 정체성을 찾을 수 없다. 건축가는 너무도 명확하게 모든 건물의 건축을 바꾸고 팽나무의 강하고 견고한 모습을 재규정하는 것을 목표로 했다.

### 21세기 철골구조의 유리상자

이러한 일은 일련의 단순한 움직임을 통해 성취했다. 첫째로는 비공식적이면서 잎과 비슷한 모양을 선택했다. 그 어디에도 이 선들을 이끄는 아날로그의 기하학적 계층이나 눈에 띄는 디지털의 알고리즘이 존재하지 않는다. 이 선들은 오래된 나무와 경관을 향한 존중의 형태를 그리는 사랑스러운 몸짓이다. 다이어그램은 프로그램 중심적이지 않으며, 삼분엽의 배치는 또렷하게 형태를 기능보다 우선순위에 둔다. 프로젝트의 주요 과제는 가능한 많은 테이블로 공간을 채우는 것이 아닌, 건축과 자연 간의 조화다. 반면에 3개의 동은 더 넓은 공간을 만들고 덜 혼잡하며 소음을 더 잘 관리할 수 있게 해준다. 여기서

특별히 두 번째로 언급할 만한 건축가의 디자인 수단은 벽이 지닌 지속적이고 투명한 본성이다. 재차 언급하지만, 건축가는 이중의 그리고 모순된 목표를 지향했다. 그는 완전한 투명성을 원하는 반면 잎을 본뜬 건물의 형태가 공간의 가변성 속에서 전적으로 사라지지 않기를 바랐다. CJ 건설(시공사)과 일진유니스코(기술사)의 야심찬 시공 과정에서 제작되었던 세련된 이중 곡선의 유리 파사드는 남다른 퍼포먼스를 제공했다. 이는 우리의 시선이 나인브릿지 파고라에서 경관으로, 주변 건물들로, 그리고 증축 부분으로 방해받지 않고 움직일 수 있도록 했다. 이 마지막 특징은 특히 커다란 파티를 하는 경우 사람들이 같은 연회에 소속되어 있는 느낌을 받아야 할 때 더욱 중요해질 것이다. 하지만 유리 파사드는 수평적인 요소들로 이루어진 매우 단단한 프레임으로 되어 있고, 잎 모양 건물의 해체된 뼈대를 재구성하여 새로운 파빌리온에 개성을 부여한다. 그리고 어떻게든 프로젝트의 종합적인 이미지를 완성하기 위한 선택은 당연히 뒤집어진 구조물이다. 이는 파사드들과 유리 지붕을 통하여 또렷하게 보여지며, 실내 공간에 최대의 개방감과 연속성을 부여한다.

건축가가 드러내는 야망과 선택의 폭을 읽고 파빌리온의 일반적인 가치를 생각해보면 몇몇 특이점들이 눈에 띌 것이다. 흥미로운 점은 이러한 주장에도 불구하고 결국 이 나인브릿지 파고라와 팩스턴이 런던에 지은 수정궁은 우리가 생각해낼 수 있었던 모든 이상의 것을 공유한다는

continuity, which characterises the indoor space.

Leaving aside for a moment the reading of the singular ambitions and choices displayed by the architects and going back to the more general values achieved by such a project, there are a few notes that could be added. It is interesting that, in the end, the pavilion shares more than we could possibly imagine with its ancient predecessor built by Paxton in London after all these argumentations. It is again a glass box with a metal structure; it is again an empty space to be filled with activity; it is again a sophisticated expression of sensibility towards the human and natural space; it is – finally – a fragment of pure innovation in form, structure and infrastructure. Once again South Korean architecture of the early 21st century proves to be lively and interesting, especially when it comes to buildings of medium/smaller scale that incorporate a sense of space, of humanity and of their landscape. As many of his generation, Lee JeongHoon succeeds in achieving an architecture that both confronts global issues – nature, technology, architecture as a small strategic intervention within the built environment – and at the same time mindful of regional specificity. Or, perhaps we should say, it transmits the spirit of the place. By choosing the hackberry tree as the main reference, this project shows its very Eastern ability to establish a dialogue with presences that are both invisible and inevitable.

An old literary passion was brought to mind when observing this project. At the beginning of my critical practice, still developing in its maturity, I was very much influenced by a beautiful and unusual text by Italo Calvino, a writer often mentioned by architects for his *Invisible Cities* (1972). My choice instead was the *Six Memos for the Next Millennium*, a beautiful book collecting the content of a series of lectures written by Calvino for Harvard University (and never delivered) and published in Italian and English versions in 1988. The next millennium has been here for a while now, and it is perhaps charming to surrender to the temptation of using the Italian master's memos as a checklist to discuss JeongHoon's architecture. For Calvino, the features to be found in a work of art in the second millennium were lightness, quickness, exactitude, visibility, multiplicity and consistency. To my view, it is surprising how perfectly the NINE BRIDGES Pergola seems to address many of these characteristics and therefore to embody the spirit of its time. Lightness, exactitude, visibility, multiplicity, could be directly attributed to this project, maybe because this is what architecture should pursue these days, or maybe because the Far Eastern cultural appreciation of space seems to be the best background to such an attitude.

사실이다. 다시 한 번 말하지만 이것은 철골 구조의 유리상자이다. 아직은 비어 있는 공간은 다양한 활동으로 채워질 것이다. 또한 인간적이고 자연스러운 공간을 향한 세련된 표현이다. 형태, 구조, 그리고 시설의 모습을 한 순수한 혁신의 조각인 것이다. 거듭 말하지만 21세기 초의 한국 건축, 특별히 공간, 인간, 그리고 경관의 감각을 포함하는 중소 크기의 건물들은 활기 넘치고 흥미롭다. 건축가는 그의 세대 대다수가 그러하듯 세계적인 이슈인 자연, 기술, 건축과 같은 지어진 환경들 이내에서 작고 전략적인 것들을 대면하고 지역적 특수함을 고려하는 데 성공했다. 혹은 달리 말하자면 장소의 의미를 생각했다고도 할 수 있다. 사실상 팽나무를 고려함으로써 이 프로젝트는 무형이면서도 필연적인 존재들과 소통하는 매우 동양적인 프로젝트가 되었다.

  이 프로젝트를 보는 와중에 나의 오래된 문학적 열정이 내 마음속에 상기되었다. 휠씬 미숙하고 비판적인 실무를 하던 나의 초기 시절, 나는 이탈로 칼비노의 매우 아름답고 특이한 글에 큰 영향을 받았었다. 많은 건축가들은 그의 저서 『보이지 않는 도시』를 언급했으나, 나는 그의 『새 천년을 위한 여섯 가지 메모』라는 하버드 강의(그리고 강연되지 않았던) 내용을 모은 책을 더욱 높이 산다(1988년 이탈리아어와 영어로 출판됨). 새천년이 임한 지는 꽤 되었고, 이 이탈리아 장인의 메모를 이정훈의 건축을 논하기 위한 체크리스트로 사용하고 싶은 유혹에 순응하는 것 역시 매력적인 일로 보인다. 칼비노는 다음 세기의 작품에서 발견되어야 할 특징으로 경량성, 민첩성, 정밀성, 가시성, 다양성 그리고 일관성을 들었다. 나에게는 나인브릿지 파고라가 완벽하게 이러한 특징들에 부합하며 그로 인해 시간의 정신을 구현하고 있는 점이 정말 놀라울 따름이다. 경량성, 정밀성, 가시성 그리고 다양성은 이 프로젝트의 직접적인 속성들이라 할 수 있다. 이는 아마도 근래의 건축이 지향해야 하는 방향이어서 일 수도 있고, 아니면 극동지역 공간의 문화가 이러한 사고방식을 형성하는 최상의 배경이기 때문일 수도 있다. 칼비노는 자신의 작업을 완성하지 못했다. 그는 안타깝게도 강연을 채 마치지 못하고 죽었고 여섯 번째 글을 쓸 시간조차도 갖지 못했다. 그래서 그의 책에는 여섯 번째 메모가 빠져 있다. 스타일, 언어, 이데올로기가 대부분이었던 1980년대에 비하여 일관성이라는 아이디어가 매우 다른 방법으로 해석될 수 있는 것도 사실이다. 오늘날의 여러 흥미로운 건축가들에 대해 말하자면, 이정훈의 건축은 되풀이하여 발생하는 이미지보다는 장소, 기술, 문화와 환경 간의 세련된 관계에 기반을 두고 있다. 이는 21세기의 일관성이라는 아이디어를 재정립하고 칼비노의 노력을 완성하기 위한 좋은 시작점이라고 할 수 있다.

fig. 5 프로젝트의 주요 과제는 가능한 많은 테이블로 공간을 채우는 것이 아닌, 건축과 자연 간의 조화다. The project's main task is to privilege the marriage between architecture (and its guests) and nature rather than filling the space with as many tables as possible.

fig. 5

Calvino could not complete his work. He died before he was able to deliver the lectures, and he didn't even have the time to write the sixth paper. Therefore the book is missing the sixth memo. It is true that the idea of consistency can be interpreted in a very different way today compared to that of the 1980s, when it was mostly about style, language, and ideology.

As for many other interesting architects today, Lee JeongHoon's architecture is not based on a recurring image but on establishing a sophisticated relationship between place, technology, culture and the environment. It could be a good starting point from which to redefine the 21st century idea of consistency and from which to build on Calvino's visionary thinking.

크리스티앙 프랑소아
낭시 건축대학
유리 디자인 학과장

# 철과 유리의 건축 역사

# History of Architecture with Steel and Glass

Christian FRANÇOIS
**Professor,
Master verre Design Architecture of
Architecture Nancy**

**과거뿐만 아니라 이정훈의 근작을 비롯한 최근 건축물에서 빛과 투명성에 대한 탐구를 찾아볼 수 있다. 이러한 건축물은 장소와 그 자연환경에 집중한 새로운 사고방식을 밀접하게 드러냈다. 이러한 양상은 건축의 문화를 바꾸면서 새 지평을 열었다.**

### 빛은 건축에서 가장 중요한 재료

루이스 칸은 이론과 실무, 교육에서 시시각각 변화하는 자연광은 건축의 공간 구조를 정의하고 나아가 건축이 탄생되는 필수적인 요소라고 생각했다. "자연 빛이 없다면 그것은 방이 아니다", "건축은 방을 만드는 데서 시작한다"라는 그의 말은 끊임없이 인용되고 있는데 방을 만드는 일이 자연광 없이는 불가능하다는 것을 의미한다.

과거뿐만 아니라 이정훈 건축가의 근작을 비롯한 최근의 건축물에서도 이러한 빛과 투명성에 대한 탐구를 찾아볼 수 있다. 우리는 흔히 중요한 역사적 기준이 되는 건축물에서는 빛이 만드는 관계, 즉 투명성과 그 투명성이 반영된 구조를 명확하게 표현하기 위한 세밀한 기법들이 직접 드러남을 볼 수 있다. 이러한 건축물들은 장소와 그 자연환경에 집중한 새로운 사고방식과 밀접하면서도 조화로운 관계를 바탕으로 지어졌는데, 이는 당대 건축 문화를 바꾸면서 새로운 지평을 열었다.

유럽의 종교건축에서 스테인드글라스를 통과해 만들어진 빛의 상징적인 의미는 바로크 시기까지 반복되던 주제였다. 이 시기의 건축은 빛의 변화가 갖는 유동성과, 그에 따라 생동감이 더해진 소재에 빛이 미치는 효과가 그 특징이다. 시간이 흐르면서 종교적 상징주의를 벗어나 건축가들은 투명한 유리와 직사광선을 기능적으로 활용하여 더 작은 규모의 나무와 유리 틀을 구성하는 데 응용했다. 이는 생물학적 조건과 원예작물의 성장과 보다 밀접하게 관련되어 있다.

### 식물원 건축의 시작

17세기 말부터 18세기 초에 이르기까지, 당시에는 신대륙 발견으로 인한 국제적인 연구와 교류에 대한 열망이 이국적인 것들에 대한 관심을 불러일으켰고, 영국은 물론 프랑스에서도 식물학의 발달에 관심이 증폭되었다. 그 당시 문화에서는 건축물을 알리기 위해 공원과 정원의 장소에 담긴 신화적 이야기와 자연의 질서를 활용했다. 건축물과 어우러진 다양한 감성의 정원 디자인에 사람들은 환호하기 시작했고, 그 속에서 식물학에 대한 새로운 관심은 커져만 갔다. 그 결과

**One can find the way that architecture from past to recent days explore light and transparency, including the recent works of Lee JeongHoon. These structures have exposed new ways of thinking that focus on the place and its natural environment. This aspect opens up a new horizon by changing the culture of architecture.**

## Sunlight is the Most Important Material in Architecture

Louis I. Kahn, according to his theoretical approach, practice and pedagogy, considered changing natural light to be a vital factor in defining the spatial structure of architecture and the conditions of its making. His famous aphorisms referring to light are frequently cited: 'A room is not a room without natural light'; 'Architecture comes from The Making of a Room', which means that, one cannot devise a room, the starting point of architecture, without natural light.

The search for light and transparency is one of the most important contexts when describing the history of architecture, not only when considering the past but also when observing recent developments, including Lee JeongHoon's recent project. The quality of light and transparency, and of transparency and detailed techniques that express structures which reflect light, has become identified with architecture that has established important historical standards. These buildings were built to create harmonious and close relationships, inforemed by new thinking that focused on place and the natural environment, changing architectural culture and opening new horizons.

The symbolic meaning of light was heightened by the prevalence of stained glass in European religious architecture up to the Baroque period. The architecture of this period is characterised by the movability of changing light and the added liveliness of its effect across the building materials. Aside from the religious symbolism, transparent glass and direct sunlight were employed in the composition of smaller wood and glass frames, which were more directly related to biological conditions and the growth of horticultural plants.

## Beginning of Botanical Garden Architecture

From the end of the seventeenth-century to the

fig. 1

조경술이 완성되었고, 다양한 건축 기법을 통해 대형 공원이 개발되고 정원조경술이 발전되었다.

르네상스에서 비롯된 이태리식 정원과 이를 재해석한 프랑스식 정원의 기하학적 구조가 유행한 후에는 자연적인 상태에 더 가까운 다른 조경 디자인이 유행하게 되었다. 채스워스에 있는 브라운Brown이나 프티 트리아농의 식물원을 만든 클로드 리샤르Claude Richards와 같은 유명한 설계자들의 이름이 새겨진 중요한 조경물들이 등장했다. 이처럼 자연에 대한 열망은 비용이 많이 드는 타지로의 탐험으로 이어졌고, 이를 통해 새로 들여온 희귀하고 이국적인 식물들의 토착화와 재배를 위한 시설물에 대한 직접적인 건축 연구가 시작되었다. 볼리비아의 아마존 수련, 난, 야자나무 등과 같은 외래식물들은 다양한 크기, 특성, 원산지 등에 따라 과학적 연구와 이해의 대상으로 여겨졌다. 파리 식물원Jardin des Plantes 원장인 뷔퐁Buffon이나 분류학을 고안한 것으로 유명한 리네Linné와 같은 뛰어난 인물들은 생물계의 질서와 다양성을 과학적으로 해석하고 알리고자 하는 열망으로 생물의 분류를 발전시켰다. 이에 따라 건축가들은 건축의 고전적 원칙을 바탕으로 구성한 건물 벽에 남쪽으로 큰 창을 내어 정원과 직접적으로 연계했다. 큰 창을 설치해 태양의 영향을 더욱 강조했고, 새로운 환경의 혹한으로부터 식물들을 보호할 수 있는 적합한 환경을 건물 내부에 갖춘 '오랑주리(오렌지 온실)'를 고안했다.

이후 조경이라는 주제는 특히 더욱 기능적인 면을 추구하면서도 시적인 감상을 잃지 않고 발전했다. 그즈음 최초의 대형 철제 온실이 등장하는데, 샤를 로오 드 플뢰리Charles Rohault de Fleury가 1833년 파리에 지은 자연사 박물관이 그것이다. 이 건물의 기술적인 혁신은 이미 알려진 사실들을 실용적으로 이해하는 데서 시작한다. 목재보다 훨씬 강도가 높은 강재를 사용함으로써 구조의 단면을 대폭 줄일 수 있었고, 강재의 휘기 쉬운 점은 건축물의 특성을 돋보이게 했다. 또한 천장에 이중 볼트를 사용한 기하학적 설계는 그림자의 영향이 크게 감소하여 직사광선의 품질과 유입되는 햇볕의 양을 늘릴 수 있었다. 우아하고 완벽하게 건설된 자연사 박물관은 확실히 기존 고전 건축 언어와는 달랐으며 대중과 과학자들에게 열렬한 환영을 받았다. 이 훌륭한 건축물과 부속 별관들 덕분에 '파리 식물원'은

fig. 1 Jean-Baptiste Hilair, Jardin du Roy. L'orangerie, 1794.
건축가들은 새로운 환경의 혹한으로부터 식물들을 보호할 수 있는 적합한 환경을 건물 내부에 조성하기 위해서 건축 내부에 더 적합한 환경을 갖춘 '오랑주리(오렌지 온실)'를 고안했다.

Architect create a more favourable microclimate to protect the plants from the heat of their new built environment, which soon came to be known as 'Orangeries'.

beginning of the eighteenth-century, the desire for international research and knowledge exchange on the discovery of the New World, still relatively recent at that time, provoked interest in the exotic and in the development of botany in France as well as in England. With this coherent and shared interest in botany, mythical stories of large parks and gardens, and an appreciation for the symbolic order of nature, revealed a range of emotions and levels of formal response according to the locations of buildings and to their corresponding culture. Landscape architecture was completed and developed through various architectural techniques according to the art of landscaped gardens and the development of large public parks.

Italian gardens originated from the Renaissance and reinterpreted in a French-style garden geometry, while other landscaping designs seeking to get closer to the natural state became fashionable. Notable garden landscapes appeared, devised by famous designers like 'Capability' Brown in Chatsworth and Claud Richard, who created the botanical garden of Petit Trianon. This craze for nature has often led to distant and costly explorations and given rise to architectural research into acclimatization and the cultivation of rare and precious exotic plants. These plants, such as the giant water lilies of Bolivia, orchids, and palm trees, were regarded as the subject of scientific study according to their various sizes, characteristics, and origins. Prominent figures such as Buffon, director of the Jardin des Plantes, and Linné, who is famous for establishing taxonomy, developed the classification of life with a desire to scientifically interpret and communicate the order and diversity of the biological world. Thus, large windows were introduced to the south walls of the buildings and designed according to the classical rules of architecture so as to establish a direct relationship to the art of garden. This amplified the influence of the sun and helped to create a more favourable microclimate to protect the plants from the heat of their new built environment, which soon came to be known as 'Orangeries'.

Some years later the theme was developed, specifically in search of more functional aspects without losing poetic sentiment, and the first large steel greenhouse appeared in form of the Museum of Natural History in Paris, built by Charles Rohault de Fleury in 1833. Its technical innovations mainly resulted from a pragmatic understanding of critical observation. The strength of steel, which is superior to wood, can greatly minimize the sections of the structure and bendability increases the properties of the building. For example, the geometric layout of double vaults increases the solar gain, and the quality of direct light, by greatly reducing the effect of shadows.

fig. 2

완전히 새로운 차원으로 도약했으며 이후 중요한 기준으로 자리 잡았다.

그로부터 14년 후, 파리 샹젤리제 거리에 지어진 겨울정원Jardin d'hiver은 이러한 도시적, 정치적인 역동성의 뒤를 이었다. 겨울정원은 샤를르 테오도르 샤르팡티에Charles Théodore Charpentier가 이폴리트 메이나디에Hyppolite Meynadier의 프로젝트를 바탕으로 설계한 것이다. 이곳은 곧 만남의 장소가 되었고, 많은 사람들이 폭 65m, 길이 100m의 대형 볼트 유리 천장을 통해 들어오는 빛을 받으며 분수와 다양한 이국적 식물 사이를 산책하거나, 행진을 하거나, 춤을 추며, 화초를 구입하기도 하고 때로는 정치에 대해 논쟁을 벌이기도 했다. 1849년에 빅토르 위고는 겨울정원에 대해 "이곳에 들어서면 처음에는 눈부신 빛 때문에 눈을 감게 된다. 이 빛을 통해서 우리는 열대와 플로리다 지역의 잎사귀를 가진 온갖 종류의 아름다운 꽃들과 기이한 나무들을 구분할 수 있다"고 했다.

그로부터 정확히 1년 뒤, 영국 채스워스

**fig. 2** Champs-Élysées. Jardin d'Hiver, 1850.

겨울정원은 샤를르 테오도르 샤르팡티에가 이폴리트 메이나디에의 프로젝트를 바탕으로 설계한 것이다.

Le Jardin d'hiver (Winter Garden) built on the Champs Elysées in Paris and designed by Charles Théodore Charpentier.

His elegant and perfectly constructed Natural History Museum was markedly different from the formal language and conventions of classical architecture and was welcomed by scientists and the general public. Thanks to this splendid work of architecture and the attached annexed galleries, the Jardin des Plantes de Paris opened up a whole new dimension to landscape architecture and has become a permanent reference point.

14 years later, another urban facility pursued a similarly dynamic urban and political approach, Le Jardin d'hiver (Winter Garden), built on the Champs Élysées in Paris and designed by Charles Théodore Charpentier. It was based on the project of Hyppolite Meynadier, and also obtained a great degree of success. Soon after its opening, it became a popular meeting place and many people used to walk under the light emanating from the large glass vault about 65m wide and 100m long, and wander between the fountains and various urbanised plants, free to roam, parade, dance, argue politics, or to even buy plants. Victor Hugo noted in a review of the Winter Garden in 1849: 'When you enter, you will close first your eyes from the dazzling light. Through this light we can distinguish all kinds of beautiful flowers and strange trees with leaves from the tropics and Florida...'

It is no coincidence that precisely one year later Joseph Paxton, an ingenious English landscape gardener who oversaw the famous Chatsworth Estate, proposed sketches for a glass and steel palace project to house the first world exhibition of 1851 in London. These designs were put together in just 8 days, informed by his previous experience of building and managing greenhouses. The Crystal Palace had to be completed in record time to retrieve the delays incurred by unsuccessful consultations and to accommodate the use of machinery and industrial products from all over the world.

Paxton's previous experience with controlling the flow of water, among other responsibilities during the creation of the Emperor Fountain in 1844 to commemorate the arrival of Nicholas I of Russia, must have played a decisive role in successfully completing this organic, ecological and futuristic building by the deadline. Before construction could begin in earnest, hydraulic energy resources, as one of the fundamental conditions for the success of the project, had to be considered, to decide how best to control a large-scale indoor air environment by supplying water to the fountain. The exemplary work performed by the Crystal Palace was considered innovative from many aspects, such as the enormity of its size, its transparency and the expression of light in spite of its scale. In years past there had

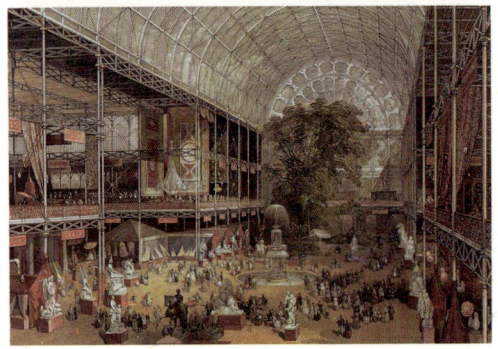

fig. 3

저택의 정원사로 유명한 조셉 팩스턴은 수정궁을 선보이게 된다. 그가 자신의 온실 건설 및 관리 경험을 바탕으로 1851년 런던에서 열린 최초의 만국대박람회를 위해 강철과 유리로된 궁전의 스케치를 8일 만에 완성하여 선보인 것은 결코 우연이 아니다. 수정궁은 잘못된 자문 때문에 지연된 사업을 만회하고 전 세계에서 온 기계들과 공업 제품들을 전시하기 위해서 짧은 시간 내에 완공해야 했다.

팩스턴은 1844년 러시아의 니콜라스 1세의 도착을 기리기 위해 '황제의 분수Emperor Fountain'를 조성하면서 맡았던 여러 업무 중에서 물을 관리했다. 이 경험은 이러한 유기적이고 생태적인 미래형 건축물을 기한 내에 성공적으로 완성하는 데 결정적인 역할을 했다. 엄밀한 의미에서 공사에 앞서 수원을 마련하는 일이 수정궁 프로젝트 성공에 중요한 조건들 중 하나였는데, 분수에 물을 공급함으로써 대대적으로 내부 공기와 환경을 조절할 수 있었다. 수정궁이라는 훌륭한 작품은 건물의 어마어마한 규모에 비해 투명하고 가볍게 표현된 점 등 많은 부분에서 혁신적이었다. 그 이전에 영국에서는 온실 건설에 대한 사회적 요구가 점점 늘어나면서 더 큰 창에 대한 연구가 늘어났고 프랑스와 영국 사이의 노하우 공유 및 기술이전을 위한 환경이 마련되었다.

특출난 재능과 경험을 갖춘 프랑스 슈아지르루아 지역의 유리제조의 대가 조르주 봉탕Georges Bontemps은 영국의 맞수인 루카스 찬스Lucas R. Chance가 작업을 발전시키는 데 도움을 주었다. 찬스의 회사는 길이가 1.2m에 이르는 보다 큰 유리관을 생산할 수 있게 되었다. 찬스의 유리공장에서 이뤄진 완벽한 유리 제조술과 1848년부터 이루어진 봉탕의 관리감독 덕분에 팩스턴의 프로젝트가 제한된 시간 내에 요구했던 생산수단과 생산량의 문제에 빠른 해답을 제공할 수 있었다. 시공 과정의 세부 사항이 합리적이었을 뿐만 아니라 건물 구성 요소의 공업화, 유리, 강철, 주철 및 홈을 판 목재 프레임 등 조립식 구조물을 사전 제작했다. 유리 8만 4,000m², 주철

fig. 3 팩스턴은 온실 건설 및 관리 경험을 바탕으로 1851년 만국대박람회를 위해 강철과 유리로된 궁전의 스케치를 8일 만에 완성했다.

Joseph Paxton, an ingenious English landscape gardener who oversaw the famous Chatsworth Estate, proposed sketches for a glass and steel palace project to house the first world exhibition of 1851 in London. ⓒMcNeven,J(Artist)/Simpson, william(Lithographer), 1851.

been a growing social demand for greenhouse construction in Great Britain, facilitating research on larger windows, and this had created the ideal environment for knowledge exchange and 'technology transfer' between France and England.

Georges Bontemps, a distinguished and talented master glassmaker of Choisy le Roi, helped his English counterpart Lucas R. Chance develop his work. As a result his company were able to produce larger sheet glass, up to 1.2m in length. Thanks to the mastery of glass manufacturing by Chance's factory, managed from 1848 by G. Bontemps, the question of the means and rate of production that Paxton's project required in limited time could be met with a quick answer. The industrialisation of building components and the prefabrication of glass, steel, cast iron and grooved timber frames, as well as the more rational details of the implementation process made it possible to carry out a project consisting of 84,000m of glass and 4,000 tons of cast iron, and built by 5,000 people in a record construction time, considering its scale, all of which remains astonishing even today. During this very short timeframe, Queen Victoria frequently visited the site, visits that numbered about 50 in total, and she adored watching the development and process of construction closely. The amplified effects of the classical typology, tightly organized composition and the arrangement of repetitive rhythms of the structure all contributed to the overall spectacle and political nature of the Crystal Palace, clearly demonstrating the performance and efficiency of industrial technology and shortening the construction period to sit within one year. The unique geometry of the design of the vaults, with three-dimensional folds like that of organic architecture, also contributed to remarkable structural strength, which ensured a great longevity to a project that was elegant, durable and-of easy disassembly.

Joseph Paxton, in his experience as a garden landscape architect, carefully observed the structure and performance of the plants, which must have served as models of design and as a vital source in his design thinking. He carefully observed a giant waterlily called 'Victoria Amazonica', which reveals how he developed his initial idea. He first managed to cultivate and encourage this remarkable plant to bloom in the greenhouse of Kew Gardens, where he analysed it with great attention to its structure and reproductive process. The iconic status of the Crystal Palace was garnered from the international scrutiny it received, and was successful from the moment of its opening. It effectively created a new archetype for architecture that is eternally engraved upon the memories of architects, as

4천 톤, 시공인원 5천 명이라는 거대한 규모에도 불구하고 오늘날의 기준으로도 기록적으로 짧은 기간 내에 시공을 마칠 수 있었다. 매우 짧은 기간 동안에 빅토리아 여왕은 50번이나 현장을 수시로 방문했고, 공사의 전개 과정에 감탄하면서 주의 깊게 지켜보았다고 한다. 반복적인 리듬으로 조직되고 정돈된 고전적인 유형의 구성이 만들어낸 효과는 수정궁이 지니는 전시적, 정치적, 총체적인 특성에 기여하면서 공업기술의 성능과 효율을 명확히 보여주었다. 동시에 1년 이내로 공사 기간을 단축해내기도 했다. 볼트 천장은 입체적인 주름들을 유기적 건축처럼 조직해 디자인 되었는데, 이 천장이 자랑하는 특별한 기하학을 통해 건축물의 구조적 강도 또한 크게 강화됐다. 이로써 우아하고 영구적인 조립식 건축물의 내구성을 대폭 끌어올려 건물의 수명을 연장했다.

조셉 팩스턴은 정원조경가로서 일하면서 식물의 구조와 그 기능을 세심하게 관찰했는데, 이것이 그의 디자인 아이디어의 원천이자 발상의 모델이 되었을 것이다. 예를 들면, 그가 자세히 관찰했던 '빅토리아 아마조니카'라는 커다란 아마존 수련에서 우리는 그가 어떻게 발상을 발전시켰는지 잘 알 수 있다. 그는 우선 큐 가든Kew Garden의 온실에서 이 특별한 식물을 키워 꽃을 피워냈는데, 이때 아마존 수련의 구조와 번식 과정에 관심을 가지고 분석했다. 수정궁이라는 상징적인 프로젝트는 개장하면서부터 바로 국제적 성공을 거두었고 새로운 건축 유형을 만들어냈다. 이는 예술, 디자인, 산업과 문화 사이의 상승작용을 불러일으킨 재능과 창조의 모델로서 건축가들의 기억 속에 영원히 새겨져 있을 것이다.

수정궁은 공사의 합리성을 여실히 보여주는데, 이는 팩스턴이 여러 분야에 걸쳐 쌓은 다양한 경험과 그의 열린 생각의 직접적인 결과물이다. 또한 이 온실은 이전의 지식과 생산 방법을 아우르는 디자인을 설계하기 위한 방법론적 중요성과 산업혁명을 향한 추진력을 보여주는 19세기의 중요한 지표가 되었다. 또한 유리 생산 및 시공 역사는 두말할 필요 없이 수정궁의 결과로 이어진 개방된 문화교류 차원에서도 매우 중요한 역할을 했다.

### 현대건축의 새로운 시도

이러한 전체 디자인에 흐르는 건축적 사고는 20세기 초 문화, 산업, 상업, 예술, 환경 등이 연계된 시너지를 바탕으로 발달한 창의적인 본보기가 되었고, 유럽에 있는 다양한 학교에서 시도되고 발전하게 된다. 그중 에콜 드 낭시Ecole de Nancy에서는 유리와 원예에 대한 여러 연구들이 이뤄졌고 일부는 제1차 세계대전 발발로 중단되기 전까지 계승되었다. 1925년

**fig. 4** 큐가든의 볼트 천장은 입체적인 주름들을 유기적 건축처럼 조직하여 디자인 되었다. 이 천장이 자랑하는 특별한 기하학을 통해 건축물의 구조적 강도 또한 크게 강화했다. The unique geometry of the design of the vaults, with three-dimensional folds like that of organic architecture, also contributed to remarkable structural strength durable and -of easy disassembly.

fig. 4

a model of ingenuity and synergetic creation situated between art, design, industry and culture.

The Crystal Palace clearly demonstrates the intelligence of the construction, which is the direct consequence of his open mindedness and transdisciplinary experience, but it also became an important indicator of nineteenth-century sensibilities. It disclosed the driving forces behind the industrial revolution and the methodological importance of the conception of a true design, one that integrates prior knowledge, experience and means of production from the conception of a work to its life beyond construction.

### A New Attempt at Contemporary Architecture

It is also an important indicator of the open transcultural exchange that resulted from the project, particularly in the history of glass production and construction. A strand of architectural thought concerning global design was developed in the early twentieth-century as a creative example of the synergy between culture, industry, commerce, the arts and the environment. This developed throughout a number of European schools, for example the École de Nancy carried out a significant programme of research on glass and horticulture, some of which was inherited in a certain way, until World War I put an end to this area of research. In the International Exhibition of Decorative Arts and Modern Industry of 1925, Peter Behrens took up the theme of a greenhouse on the bank of the Seine. A geometric volume, a rectangular parallelepiped suspended by a steel structure on the river, was composed of an elegant diagonal network of transparent glasses that accentuated its presence. Its very bright interior contrasted with the geometry and technology of its envelope, and the theme of the natural world was noted in its rustic character. With an exhibition space resplendent with plants, fountains, stones and spherical luminaires, Behrens proposed a project that sought a synthesis between

fig. 5

현대장식미술·산업미술 국제박람회Exposition internationale des Arts décoratifs et industriels modernes에서 페터 베렌스는 센 강변의 온실이라는 주제를 다시 이어갔다. 강가에 설치된 기하학적 대형 건물은 철골구조에 매달린 직육면체 형태로 그 존재감을 강조하는 우아한 유리 사선망으로 이루어져 있다. 아주 밝은 내부 공간은 외피의 기하학 형태와 여기에 사용된 기술과 대조를 이루고 있는데, 소박하게 재구성된 자연이라는 주제를 전원적인 특징으로 찾아볼 수 있다. 많은 나무와 화분, 암석, 구형 조명기구 등을 전시하면서 베렌스는 당시의 기술과 양식적 연구, 살아있는 자연을 아우르고자 하는 프로젝트를 제안했다.

문화·산업·기술·개념의 원동력은 국제교류에 적합한 덕에 다양한 분야의 지식을 넘나들며, 인간과 자연환경, 그리고 인간의 사회생활 환경의 역사 사이의 조화를 중시하는 사상과 연관된다. 이러한 원동력은 종종 다음 세기에 구현된다. 영국 기술자인 피터 라이스Peter Rice가 설계한 파리 라 빌레트 과학산업관의 유명한 온실은 20세기의 대표적인 사례로 볼 수 있다. 숲이 우거진 공원에 자리 잡은 온실의 투명성에서 영감을 받은 그의 작품은 유리를 외부에서 철물로 고정하는 새로운 구조적 설치 방식을 고안해냈다. 가는 스테인리스 스틸로 된 십자형 고정대는 뒷면에 풍압을 견디기 위해

fig. 5 데이비스 알파인 하우스는 역동적인 표현과 형태적인 해답을 새롭게 제시했다. 또한 환경의 기후조건 및 생태조건에 도움이 되는 기술에 관심을 불러일으켰다. Davies Alpine House provided a dynamic expression and morphological solutions to the architectural issues of the present, and prompted interest in technology that would aid climate control and monitor the ecological conditions of the simulated environment. Photographer stephen boisvert

contemporary technology, style and living nature.

The cultural, industrial, technological, and conceptual driving forces, crossing various fields and favouring international exchange, have been related to ideas that emphasise the harmony of the human world with our natural environment. It also considers the history of human social living conditions, and which is often found material reality in the future. The famous greenhouse built by English engineer Peter Rice at Parc de la Vilette is a good example of this architectural strand in the twentieth-century. This programme inspired by the transparency of greenhouses in public parks helped to invent a new concept for the structural implementation of glass: the structural glass façade. The fine cruciform stainless-steel fasteners are connected to the crisscrossed cables in the background, which can endure the potential bending forces of the wind. The large glass panels are suspended by a four-way spider without stress, remotely connected to the steel structure by finely designed connectors: the four-way spiders or glass fins became emblematic thereafter. This allowed a work of over 12m tall to be enriched by a high transparency and perfect flatness, which closely optically related the building to the park. Later, this solution became a structural model, followed by many variants, and the elegant volume of structural glass freed from its frame became a representative work of modernity.

In the early years of the twenty-first century, the WilkinsonEyre Architects built another greenhouse in Kew Gardens, London, and the building is remarkable for its close ties to the tradition of collecting English plants and for its stylised modern architecture. This 10m-high greenhouse, the Davies Alpine House, built in 2006, has a curved ladder of a compressed steel arch which creates a transparent roof like two curved sails made of 12mm thick ultra-transparent low iron glass. Held by thin cables, it hangs taut in the air. A protective layer permeable by vision and light made it possible to recreate the environment vital for the plants of this Alpine collection. Sunny, dry, cold, windy, and shaded conditions could be obtained by its passive natural ventilation. The building provided a dynamic expression and morphological solutions to the architectural issues of the present, and prompted interest in technology that would aid climate control and monitor the ecological conditions of the simulated environment.

Another approach can be observed in the designs of the Heatherwick Studio, who proposed a variation on the contemporary transparent greenhouse in their 2014 conversion project of the former Laverstoke Paper Mill into the Bombay Sapphire Distillery. Two twin and complementary

엮인 케이블에 연결되어 있다. 대형 유리판은 네 부분으로 구성된 모듈에 응력을 받지 않은 채로 매달려 있으며, 떨어져 있는 철구조물에 섬세하게 설계된 연결부로 이어져 있다. 이후에 이 연결부는 상징적인 존재가 된다. 이렇게 함으로써 높이가 12m가 넘는 건물에 뛰어난 투명성과 완벽한 평면성을 부여했으며, 건물과 공원이 시각적으로 밀접하게 연계될 수 있도록 했다. 이러한 건축 방식은 그 후 다양한 방식으로 변형되어 재해석되고 프레임에서 벗어난 대형 유리 건물의 우아한 고정 방법을 탄생시킨 구조적 모델이 되었고, 근대성을 보여주는 대표적인 작품이 되었다.

21세기 초 무렵인 2006년, 윌킨슨에어WilkinsonEyre건축사무소에서는 런던의 큐 가든에 또 다른 온실을 지었는데, 이 온실은 영국의 식물 채집 분야의 전통과 현대건축 양식에 밀접하게 연관된 점에서 주목할 만한 뛰어난 건축물이다. '데이비스 알파인 하우스Davies Alpine House'라는 이름의 이 온실은 높이가 10m에 달하며 압축된 강철 아치 형태로 휜 사다리 모양을 하고 있는데, 강철 아치는 가는 케이블을 이용해 두께 12mm의 극도로 투명한 저철분유리로 만들어진 투명한 외피를 공중에 팽팽하게 매달아 마치 두 개의 돛처럼 생긴 곡면을 만들어냈다. 주변 조망과 빛을 매우 잘 투과하는 유리 보호막 덕분에 온실에 있는 식물들의 생육에 필수적인 햇볕이 잘 드는 동시에 패시브 자연 환기 방식을 통해 건조하고 서늘하고 바람이 불고 그늘이 있는 환경을 재현할 수 있다. 데이비스 알파인 하우스는 우리 시대의 건축적으로 중요한 문제들에 대해 역동적인 표현과 형태적인 해답을 새롭게 제시했다. 또한 환경의 기후조건 및 생태조건에 도움이 되는 기술에 관심을 불러일으켰다.

또 다른 접근 방식으로는 2014년 헤더윅 스튜디오Heatherwick Studio가 봄베이 사파이어 사의 증류소 건설을 위한 레이버스토크의 제지공장 재개발 프로젝트에 투명한 온실이라는 주제로 새로운 변형안을 제안한 것이 있다. 2개의 쌍둥이 건물은 하나는 따뜻한 온실, 다른 하나는 차가운 온실로 상호보완적 성격을 띠고 있어 독주의 일종인 진을 만드는 데 필요한 여러 종류의 식물들을 재배하고 전시할 수 있다.

두 개의 온실은 조셉 팩스턴이 채스워스에서 처음 실현했던 전설적인 설계안을 떠올리게 하는 데 부족함이 없다. 두 건물의 외피 뼈대는 오래된 붉은 벽돌 공장이었다가 복원되어 지금은 증류 공장과 리셉션 장소 및 매우 전형적인 시음 공간으로 탈바꿈한 건물의 창턱에 역동적인 모습의 증류기처럼 연결되어 있다. 이와 같이 2개의 온실을 극적인 방식으로 연결한 것은 단순히 상징적인 표현을 위한 것만은 아니다. 양조장에서 발생한 열을 재활용하여 온실 내부를 환기하고 에너지 효율을 높이기 위한 것이다.

fig. 6 2개의 쌍둥이 건물은 하나는 따뜻한 온실, 다른 하나는 차가운 온실로 상호보완적 성격을 띠고 있다. Twin of two building and complementary buildings, a warm greenhouse and cold one, had been designed to grow and display the various species of plants used as ingredients for their gin. Photographers Hufton + Crow

fig. 6

buildings, a warm greenhouse and cold one, had been designed to grow and display the various species of plants used as ingredients for their gin. The structural design of the two greenhouses evokes the legacy of Joseph Paxton's first projects at Chatsworth. The ribs of their envelopes are dynamically connected, like a distiller, to the windows of the old restored red brick factory, which has now been transformed into a distillery, reception and attractive tasting spaces.

They are connected in a spectacular way, not merely as a symbolic gesture, but in order to recycle the heat of the distillery and to ensure the ventilation of the greenhouses, demonstrating a smart and efficient use of energy. The extension of large glass and steel volumes is in harmony with the river by matching the floor level to the surface of the water, another feature associated with the themes of environment, glass, water and light. The river, which was reconnected to the building and exhibits an abundant flow, can be also interpreted as a poetic metaphor for living, with luminous streams of water flowing back onto the floors of the greenhouses. These complementary buildings welcome people, reminding them of the close link between gin and its culture, with the deepened quality of a natural environment offering a memorable experience.

유리와 강철로 만들어진 이 2개의 거대한 건물을 하나로 구성함에 있어 환경, 유리, 물, 빛의 흔한 결합 중 또 다른 불변 요소인 강과의 조화를 고려했다. 건물 주변으로 이어진 강은 풍부하게 흐르고, 온실 바닥으로 다시 흘러 들어오는 살아 움직이는 빛나는 물에 대한 시적인 은유로 해석되기도 하다. 상호보완적인 2개의 온실은 그곳을 방문한 사람들에게 기억에 남을 만한 경험 속에서 진과 문화, 자연환경의 우수함 간의 관계를 떠올리게 해준다.

나인브릿지 파고라는 이러한 건축적 연구의 연장선상에 놓여 있다. 규칙적 환기 시스템을 통한 내부 공기 조절 공법, 최신 기술로 만든 이중 곡률을 갖는 유리의 본질, 건축적으로 완전히 투명한 이 프로젝트가 공업적으로 국제적인 협업을 통해 진행된다는 점을 충분히 고려해야 한다. 정밀한 시공 설계와 같은 사실들은 과거의 유리 건축에서 보여준 퀄리티를 떠올리게 하는 데 모자람이 없다. 형식적인 외관 외에도 이 특별한 프로젝트가 다른 해석들보다 더 중요하게 새로운 문제를 제기하는 점은 어쩌면 그 근본적인 발상에 있을 것이다. 그것은 600여 년은 족히 된 고목에 신성한 상징적인 가치를 부여하며 온전히 본래 자리에 그대로 둔 채 보호함으로써 그 모습을 계속 볼 수 있게 해주는 것이다. 어쩌면 이 프로젝트를 과거, 특히 18세기의 식물 애호가들부터 가장 최근의 사례에까지 빗대어 볼 수도 있다. 그러나 이번에는 건축이 나무의 자리를 보호하면서 그곳으로 옮겨져 투명하게 연출된 것처럼 보인다.

거대한 성목을 옮겨 심어 수용했던 수정궁의 전시회 이미지와 기존의 신비로운 나무의 자리와 그 나무가 지켜온 생명의 공간에 건축물을 섬세하게 배치한 나인브릿지 파고라를 비교하기 위해 두 건축물의 시각적인 인상으로부터 한 발 물러서서 보면 두 건축물의 접근 방식에 대한 차이는 훨씬 충격적이다. 식물학자들과 온실의 중요한 역사적 기준이 되는 건축물에서 건설자들은 외래종의 머나먼 흙을 가지고 와서 그들이 만든 투명한 온실에 옮겨 심었는데, 어떤 의미에서 이는 외래종의 본질과 역사를 존중하고 거기에 맞추기보다는 가져온 것을 자기네 것으로 만들려는 의도였다.

이와 같은 완강한 정복 의지는 최신의 지적·과학적 지식을 모두에게 알리고자 하는 계몽주의가 갖고 있는 특성이기도 하다. 이는 분명히 생물다양성과 관련된 수많은 교류의 포문을 연 것은 사실이지만 그 교류가 완벽하게 성공한 것은 아니었다. 오늘날 이 같은 과학적인 지식은 소중한 생태 및 문화 유산을 보존하는 것에 대한 완전히 공유된 인식, 즉 우리의 땅과 생활 환경을 존중하는 인식을 불러일으켜야 한다.

세계화원 산업과 우리 사회의 활동들이 야기한 새로운 문제들을 고려할 때, 지속가능한 개발 및 자원 생태학이라는 경향과 환경의 조화를 이루는 건축은 탐색과 발전, 새로운 교육적 기준을 제시할 것이라는 희망을 불러일으킨다.

The remarkable the NINE BRIDGES Pergola seems to continue this present trend in architectural research. Considering the internal air-control system is controlled by the regular ventilation system, the quality of double-curvature glass made according to the latest technology, and the precise implementation of a project that is architecturally transparent (carried out through a process of international industrial collaboration) there are no shortage of reminders here of the quality realised by the glass architecture of the past. Aside from its formal appearance, this special project advances a fundamental philosophical appreciation of nature that is of greater importance to architecture than other issues at hand. It is as if a tree of over 600 years old, endowed with sacred symbolic value, preserves and protects its site of origin so that it can be appreciated over time. This project may recall the botany enthusiasts of the past, especially those of the eighteenth-th century or even perhaps more recent cases. However, this time it is architecture that appears to be transported to the site of the tree, ordered to protect it transparently on the stage.

The difference in approach here is even more striking if we step back from the visual impressions of two projects to compare it with the exhibition image of the Crystal Palace, which housed a monumental replanted mature tree.

In the NINE BRIDGES Pergola of the present day, which delicately arranges the buildings in the space of the sacred tree, it is the tree's vital space that is preserved. The botanists and the greenhouse builders introduced the distant soil of the exogenous species and planted the plants in a transparent greenhouse, which in a sense was intended to appropriate the imported, rather than to respect and adapt to their essence and history.

This strong will to conquer is a guiding feature of the Enlightenment, which announced the latest intellectual advancements and breakthroughs in scientific knowledge, and encouraged access to these findings for all. It is true to say that the period initiated many significant exchanges on biodiversity, but these efforts were not entirely successful. Today, the sharing of scientific knowledge should be advanced by a greater awareness of the preservation of our precious ecological and cultural heritage, and reinforcing respect for and the perception of our land and living environment.

Considering the new challenges presented by globalised industry and the increased activity of our societies, new trends in sustainable development, resource ecology and a pursuit of architecture in harmony with the environment inspire hope for a new age of exploration and development, proposing new educational standards.

Photo Essay

# Pergola
### of The Club at NINE BRIDGES

**Construction Essay**

# Pergola
### of The Club at NINE BRIDGES

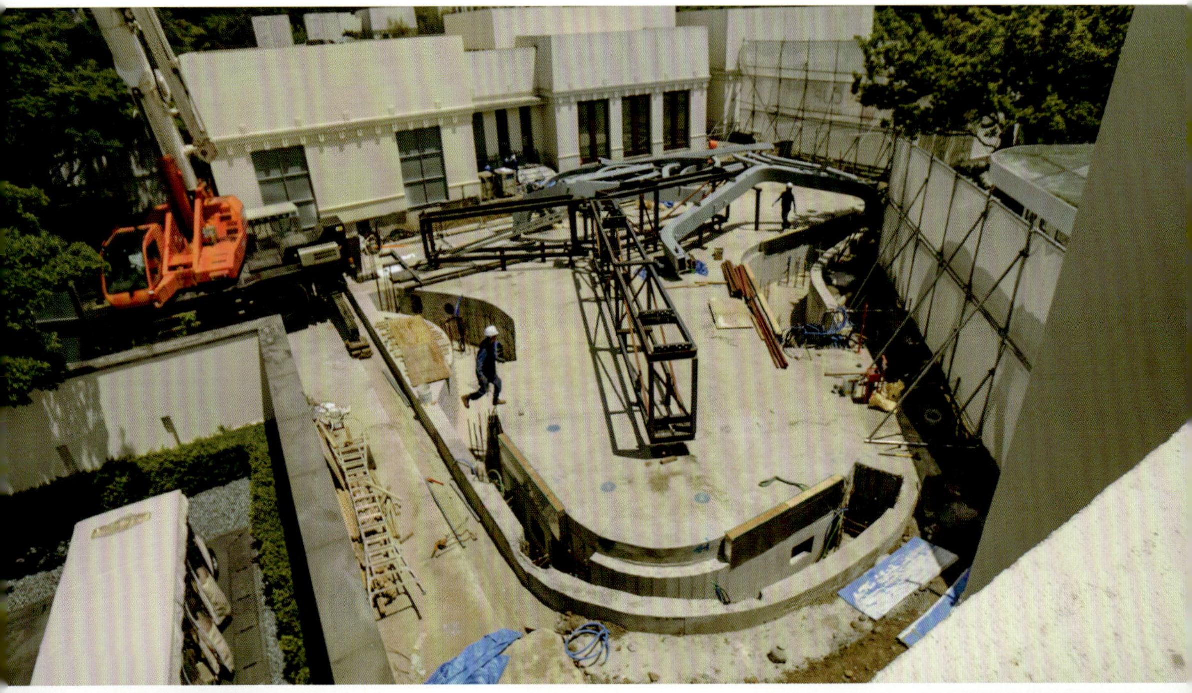

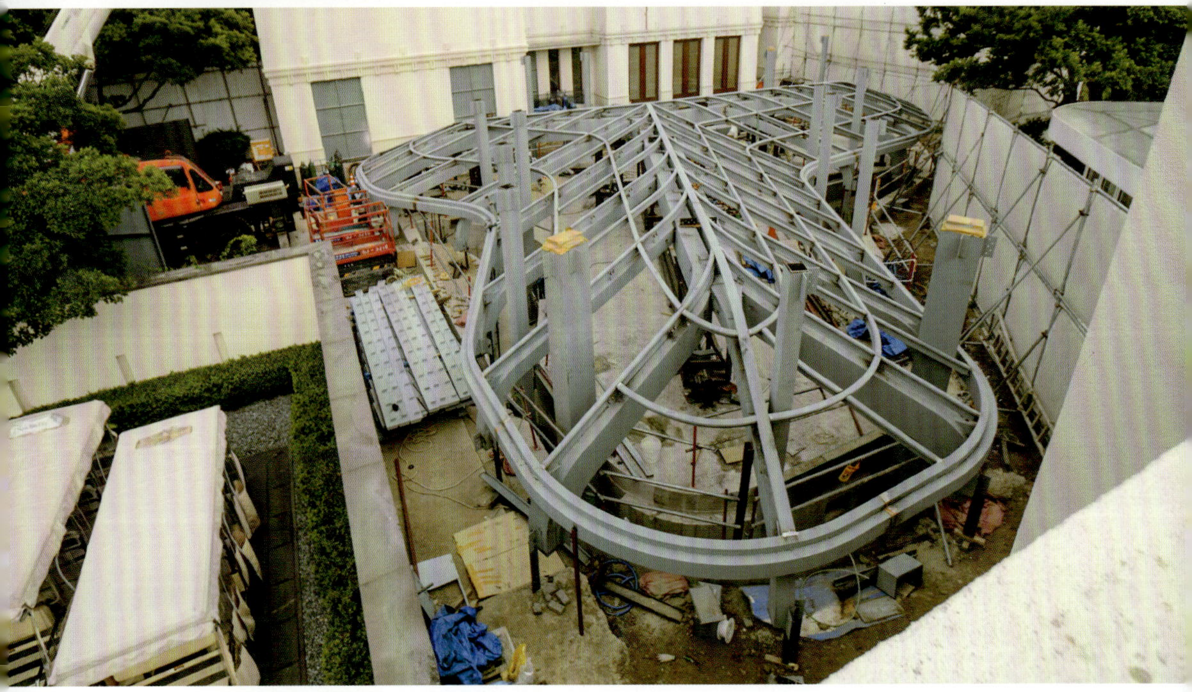

일진유니스코

# 디지털을 통한
# 건축구조와 혁신

# Architectural Structure and Innovation through Digital

**ILJIN Unisco**

클럽나인브릿지 파고라는 형태·구조·기능이 결합하여 하나의 언어로 풀어낸 시도이다.
카티아를 사용해 1만여 개에 달하는 부재에 대해 체계적으로 관리했고,
유리 시공의 효율을 높이기 위해 유리 최적화 작업,
철골 구조체 분류 작업 등 다양한 데이터를 활용했다.

### 건축의 미래 – 디지털 디자인의 시작

불과 10년 전 '디지털 디자인'의 개념이 처음 등장했을 때, 많은 사람들은 디지털 디자인이 건축 설계를 위한 '도구의 발전'이라고 간주했다. 일부 건축가는 단순히 모형 제작에만 사용했고 디지털 디자인과 건축 설계를 결부하는 일은 여전히 생소하다. 그러나 현재 디지털 디자인은 생물학적 형태를 차용하여, 3차원으로 환원하기 어려운 디자인을 물리적 실재로 구현해내는 '디지털 디자인만이 할 수 있는 건축'에 도전하며, 단순히 편리한 도구를 넘어 창작의 매개로 발전하고 있다. 즉 '표현의 도구'에서 '구축의 도구'로 변화하고 있는 것이다.

비정형 건축물을 설계할 때는 설계 오류의 최소화, 부재의 정밀한 제작 및 현장 조립이 항상 고민되는 주제이다. 일진유니스코는 특히 정밀한 3D 형상 제작 방법을 연구하는 데 시간과 비용을 투자했고, 그 결과 DDP, CJ R&D 센터, GT 타워, 전경련 회관과 같은 독특한 형태의 건축물을 성공적으로 시공했다. 특히 우리는 커튼월을 전문적으로 설계·시공해왔으며, 디지털 건축과 구현부분에서 독보적인 위치를 차지하고 있다. 나인브릿지 파고라는 커튼월 외에 비정형 구조체로 이루어져 정밀한 제작을 요하므로, 그간의 집약된 기술력을 선보이기 위해 노력했다. 나인브릿지 파고라를 진행하면서, 우리는 한 단계 더 진보한 건축에 대해서 생각하게 되었고 '다양한 디자인 방법론을 통해 건축가의 디자인을 오차 없이 실현할 수 있는가?'에 대한 답을 찾으려 했다.

디지털 디자인이라는 개념이 활발히 논의된 지 10년이 조금 넘은 지금, 국내 상황을 살펴보면

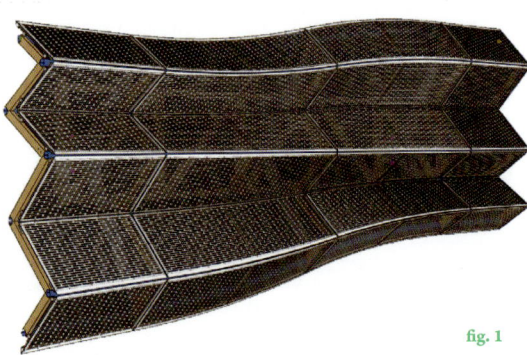

fig. 1

fig. 1 우리는 정밀한 3D 형상 제작 방법을 연구하는 데 시간과 비용을 투자했다. CJ R&D 센터는 3차원 형상의 철골 제작과 이를 기반으로 복잡한 입면 구조체를 모듈화 했다.

We spent the most significant proportion of our time and money on studying the precise methods of manufacturing a 3D shape. The CJ R&D Center conducted a 3D shaped steel fabrication and made modular of complexity elevation structure based on it.

The Pergola of The Club at NINEBRIDGES shows an attempt to combine shapes, structure and functions in one language. CATIA was used to systematically manage over 10,000 members and used various data such as glass optimization and steel structure classification to increase efficiency of glass construction.

### The Future of Architecture
### – The Beginnings of Digital Design

Only 10 years ago, when the concept of 'digital design' first emerged, many people considered it 'a development of the tool' in architectural design. Some architects used it simply for model making, and yet still today it remains relatively unfamiliar to link digital design to architectural design. Nevertheless, the current stage of progress reached by digital design is that of an energetic move towards 'an architecture that digital design can only achieve': it derives its shape from the biological form and it realises complex designs (which are hard even to transform into 2-dimensional forms) into their physical realities. Going beyond that of a simple and convenient tool for design, digital design is becoming a medium for creation. In other words, it has transformed 'a tool for expression' into 'a tool for creation'.

When designing an irregularly shaped building, certain things always become issues, such as minimising errors in design, the precise manufacture of the construction components, and their on-site assembly. We spent the most significant proportion of our time and money on studying the precise methods of manufacturing a 3D shape. As a result, we have successfully constructed buildings with unique shapes, such as the DDP, CJ R&D Center, GT Tower, and the Federation of Korean Industries in Yeouido. We have professionally designed and constructed curtain wall systems, attaining a dominant position in the field of their realisation and in digital architecture. The NINE BRIDGES Pergola was composed of an irregular structure aside from its curtain walls, requiring a precise degree of manufacturing. Therefore, we struggled to express the integrated technology devised for this project. While we were working our way up to the NINE BRIDGES Pergola, we thought about an architectural work that would take our vision one step further. We also tried to find an answer to the following question: 'by using various design

디지털 건축이 교육 현장에 도입된 지 몇 년 되지 않았고, 그렇다 보니 건축 디자인과 결부하는 작업이 그리 많지 않았다. 그러나 DDP와 전경련 회관을 시공하면서 얻은 사실은 정형화하기 어려운 수많은 부재들을 정해진 시간 내에 생산하기 위해서는 지금까지와 다른 방법론이 필요하다는 것이었다.

처음 디지털 디자인이라는 요소를 DDP에 적용하고 전경련 회관을 거쳐 지금에 이르기까지 디지털 프로그램의 건축적 연구와 개발을 지속하면서, 건축 설계안과 물리적 구축의 관계를 재편해왔다. 형태·구조·기능을 하나의 언어로 풀어내 컴퓨터상에 모델링으로 결과를 도출하고, 한 단계 더 나아가 제작에 관한 실험을 거듭하여 시공까지 이어갔다. 모두가 디지털 디자인의 개념이 제작과 동떨어져 있다고 생각할 때, 우리는 50년간 쌓아온 수공예적 제작 기술을 바탕으로 디지털을 제작 도구로 사용함으로써 우리의 기술력을 한 단계 높였다. 현재 우리의 디지털 기술의 핵심은 누구나 할 수 있는 정교한 모델링을 넘어서, 더욱더 섬세하고, 명쾌하며 목적에 맞는 디자인을 실제 형상으로 드러낼 수 있는 능력이다.

나인브릿지 파고라는 건축가가 '나무'에서 형태를 차용해왔고, 뿌리에서 줄기로 양분을 공급한다는 의미를 기능적으로 풀어나갔다. 건축의 구조체는 형상을 따라가며 그 안에 기계·소방·전기·통신 설비를 모두 삽입하여 기능과 형태를 같은 언어로 풀어냈기 때문에, 구조체와 MEP¹를 함께 해결하는 방법이 필요했다. 앞선 다른 프로젝트 때보다 모듈화에 대한 차원 높은 방법론을 요구했다.

따라서 카티아CATIA를 통해 디자인부터 제작, 그리고 각기 다른 부재 조립과 시공 계획을 세웠다. 이는 건축적 형태를 디자인하는 것뿐만 아니라 제작, 품질 개선 등을 거쳐 수많은 부재를 정해진 시간에 현장으로 운송하고 시공하는 것까지를 의미한다.

작업의 과정은 다음과 같다. 우선 카티아를 활용하여 MEPMechanical Electronic Plumbing, 기계, 전기, 소방을 통합한 모델링을 제작한다. MEP를 통합한 모델링의 강점은 비정형의 성질이 강한 건축물에서 발생하는 구조 및 MEP 도면 간의 불일치 현상을 3D로 쉽게 바로잡을 수 있다는 것이다. 구조체 안에 들어가는 설비를 비롯한 MEP 도면을 모델링 초기 단계에서부터 포함하여 설계의 오류를 최소화하면, 이후에 발생하는 도면 간의 불일치 현상을 줄일 수 있다.

모양이 제각각인 유리 및 마감재의 설계, 도면화, 제작, 시공 전 과정도 디지털 설계를 기반으로 이루어졌다. 유리 시공의 효율을 높이기 위해 유리 최적화 작업, 유리를 고정하기 위한 프레임의 최적화 작업, 철골구조체 지지체 분류 작업 등 보이지 않는 부분의 데이터를 만들었다. 나인브릿지 파고라가 국내에서 처음으로

**fig. 2** 나인브릿지 파고라는 국내에서 처음으로 형태·구조·기능이 결합한 부재를 선보였다. 카티아를 통해 디자인부터 제작, 그리고 각기 다른 부재 조립과 시공계획을 세웠다. Since the NINE BRIDGES Pergola was the first attempt made in Korea to integrate three essential elements – form, structure, and function – and express them in one language. The overall process of design, manufacture, and assembly of each different construction component, as well as the planning for construction, were all done in CATIA.

methodologies, is it possible for an architect's design to be realised without error?'

Since the concept of digital design has been actively pursued and discussed for little more than 10 years, and digital architecture was introduced to the field of education in Korea only a few years ago. There were not much work which link digital design with architectural design. However, by observing the numerous constructions of DDP and the Federation of Korean Industries in Yeouido, we learned an important lesson: in order to manufacture lots of complex construction parts that are hard to shape within a given time period, a completely different methodology is necessary.

Throughout the process of applying digital design—first in DDP, then in the Federation of Korean Industries, and now to this day—we have conducted a constant programme of architectural research and invested in the development of digital programmes. We also restructured the relationship between architectural design and a physical construction. We integrated form · structure · function into one type of language, extracted the result via computer modeling, and took further steps to experiment with the manufacturing process, which led to the actual construction. When everyone considered the concept of digital design set apart from manufacturing, we based our thinking on 50-years of cultivating our

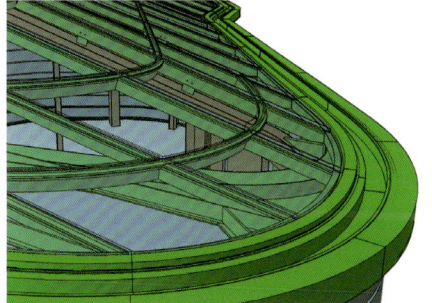

fig. 2

manufacturing techniques and utilized digital as the means to manufacture and elevate our use of technology further. Currently, the core element of our digital technology is to be able to build a form that is much more delicate, clear and purpose-oriented, one that will go far beyond the production of a detailed modelling process as found in other practices.

For the NINE BRIDGES Pergola, the architect derived the form of the building from 'a tree', and conveyed the functional implication of nutrition supply by depicting the root to the stem: in order to express the function and the form of the building in the same language, the structure of

형태·구조·기능 부분의 세 가지 필수 요소를 결합하여 하나의 언어로 풀어내려는 시도인 만큼, 1만여 개에 해당하는 부재에 대한 체계적인 관리가 필요했다.

나인브릿지 파고라의 강점은 어느 곳에서나 분해하고 이동 후 조립하여도 완벽한 조립이 가능한 정교함에 있다. 그동안 진행되었던 국내 비정형 프로젝트들을 살펴보면 철골의 경우 10mm, 유리의 경우 3mm 이내의 오차 기준을 갖고 있으나, 우리는 기존의 오차 기준을 더 줄여 시공의 정교함을 높였다. 또한 기존의 비정형 유리를 사용한 프로젝트의 기술적인 상태를 살펴보면 강화 과정을 거친 단일 유리가 단순히 치장재로 사용되는 데 그쳤다는 것을 알 수 있었다. 나인브릿지 파고라의 유리는 전량 양면 반강화 과정을 거친 3D 형태의 로이 복층 접합 유리를 사용했고, 기능적인 목적을 담고 있기에 '다음 세대의 유리'라 불릴 정도의 진화된 기술력을 보여주었다.

### 나인브릿지 파고라와 디지털 기술

나인브릿지 파고라는 시공이 진행되기 이전에 3단계에 걸친 작업이 진행되었다. 먼저 디지털 프로그램을 활용한 디자인 방법을 도출하고, 두 번째는 3D 측량기로 부분을 측량하고, 마지막으로 철골 프레임을 생산하고 검측하는 과정을 사전에 거쳤다.

fig. 3

### 카티아를 통한 디자인

앞서 설명대로 이번 프로젝트는 카티아 프로그램을 사용했다. 사용자를 위한 기능적인 목적을 하는 영역served space과 구조 및 설비들을 포함한 영역servant space이 결합되어 건물의 형태에 따라 기능적인 부분들을 배치했다. 이때 시공에 필요한 모든 도면이 완벽히 일치해야 현장에서 오류를 최소화할 수 있다. 먼저 파고라의 벽체·지붕·구조체의 데이터를 분석하고 입력하는 단계인 2D 도면을 분석하는 것에서부터 시작했다. 완성된 2D 도면의 정보를 라이노Rhino 프로그램에 입력해 기본적인 모델링을 완성한 후, 카티아로 복잡한 디테일을 마감하는 과정을 거쳤다.

이 과정에서 MEP 도면도 함께 3차원 변환

**fig. 3** 구조체는 형상을 따라가며 그 안에 기계·소방·전기·통신 설비를 모두 삽입하여 기능과 형태를 같은 언어로 풀어냈기 때문에, 구조체와 MEP를 함께 해결하는 방법이 필요했다.
In order to express the function and the form of the building in the same language, the structure of the building followed its form and its mechanical fire protection and electrical communication system were inserted into the structure. Thus, the structure and MEP had to be resolved concomitantly.

the building followed its form and its mechanical fire protection and electrical communication system were inserted into the structure. Thus, the structure and MEP[1] had to be resolved concomitantly. A well-advanced methodology on modularization was also required to a greater extent compared to previous projects.

Therefore, the overall process of design, manufacture, and assembly of each different construction component, as well as the planning for construction, were all done in catia. This does not only implicate designing of an architectural form but also includes manufacturing, improving quality, delivering numerous construction parts to the site and building them properly within the given time.

We used catia to produce MEP-integrated modeling. The strength of the MEP-integrated modeling lies with the ease of correcting any misalignment between the structural and the MEP drawings, which could arise in an irregular building with a powerful character. If all the MEP drawings – including the building system for the structure – are included at the initial stage of modeling and subsequently minimize the error in design, then a possible misalignment between drawings could be diminished.

The procedures, including design, drawing, and manufacturing of various-shaped glass components, and all the coverings, including panels and other like casings, were also based on digital designs. Data for few unseen processes were made, such as optimizing the glass to increase the efficiency of glass construction, optimizing the frames to hold glass panels, and categorizing the steel structure and support. Since the NINE BRIDGES Pergola was the first attempt made in Korea to integrate three essential elements – form, structure, and function – and express them in one language, the systematic management of almost 10,000 construction members became necessary.

The strength of the NINE BRIDGES Pergola was its ability to be perfectly reassembled and relocated on any site after being dismantled. When irregularly shaped building projects in Korea were placed under closer scrutiny, the standard (manufacturing tolerance) for the steel frame and glass were within 10mm and 3mm respectively. Yet, we reduced the existing margin of error even more to increase the precision during construction. Moreover, when we looked at the technical state of the previous projects that had used irregularly shaped glass, a single pane of glass that had undergone the reinforcing was simply used as decoration. The glass of the NINE BRIDGES Pergola was 3D shaped, low-emission, laminated tempered glass which had been put through a double-sided semi-tempering process. Since this

과정이 필요하다. MEP의 핵심 요소인 덕트의 디자인과 제작 방법이 이 과정에서 결정되었다. 구조의 역할을 하는 철골 안에 기능의 역할을 하는 덕트가 삽입되므로, 두 핵심 부재 간의 결합 방식도 정해졌다. 또한 후에 시공의 효율성을 위해 어떤 부분에서 절단을 할지, 절단한 부재를 현장으로 운반해 조립할 때 오차가 생길 경우 어떻게 수정을 할지에 대해 고민했다. 현장에서 실제로 시공을 했을 때, 더 시공 효율성이 높은 방법에 맞추어 모델링을 수정하고, 그에 따라 2D 도면을 역으로 수정하는 작업을 반복했다. 일련의 도면 분석과 변환 과정을 마쳐도 부재들이 무질서하게 나열된 상태이기 때문에, 일정한 규칙을 만들어 부재들의 순서와 절대적인 위치를 지정해주어야, 이후에 모델링을 수정하기 쉽고 시공을 할 때 적재적소에 맞는 부재를 배치할 수 있었다. 또한 이 과정에서 생성된 모델링의 결과는 대략적인 위치와 형태를 판단하기 위한 값이었다면, 다음 과정들을 통해 생산에 직접적으로 쓰일 수 있는 수준의 완성도를 가진 모델링을 완성했다.

나인브릿지 파고라는 디지털 설계의 핵심이라고 할 수 있는 지오메트리Geometry 구성 과정을 거쳤다. 이 과정은 건물의 뼈대가 되는 모든 부재를 관리할 수 있도록 하는 네이밍 작업으로, 이 프로젝트의 경우 6개의 주 철골(하중의 대부분을 지지), 19개의 부 철골, 23개의 멀리언, 12열의 트랜섬으로 구성된다. 이 뼈대의 이름은 '마스터 와이어프레임Master Wireframe'으로 문자 그대로 변하지 않는 '기준 골격'이란 뜻이며, 앞으로 추가될 부재들의 기준이 된다. 기준을 잡는 작업을 통해 후에 첨가될 구조체·마감재의 위치도 자동으로 지정되고, 데이터 값으로 빠르게 접근해 세부 사항의 추가 및 수정이 가능하므로 필수적인 작업이다.

이번 프로젝트는 설계 진행 중 제작과 시공 방법까지 고민하며 정밀한 설계안이 필요했다. 기본적인 뼈대 구성 작업이 완료되면, 살이 되는 부재들에 대한 모델링을 진행했다. 이것은 부재의 최종 형태와 마감 방식, 부재들이 만나면서 조립되는 방식 등을 모두 포함하는 자료를 말하며, 제작 및 시공에 직접적으로 사용할 수 있는 수준의 데이터를 의미한다. 우리는 뼈대에서부터 마감재에 이르기까지 현장에서의 시공 시간을 최소화하기 위해 조립 방식[2]을 채택했으며, 부재의 분할을 최소화했다.

나인브릿지 파고라는 구조체 안에 설비 기능이 삽입된 최초의 디자인이다. 12mm의 두꺼운 주 철골 안에 2mm의 덕트를 철판 형태로 유지한 채 삽입하기 위해선 결합 방법에 대한 연구와 그에 따른 정밀한 기술력이 필요했다. 철골 중 직선 부재는 삽입이 쉽기 때문에 최대한 직선 부재의 비율을 높이고, 직선과 곡선이 만나는 지점에서 300mm 떨어진 곳을 절점으로 선정했다. 곡선이 끝나는 부분에서 주 철골을

fig. 4 지오메트리 구성 과정은 건물의 뼈대가 되는 모든 부재를 관리할 수 있도록 네이밍하는 작업이다. 이 프로젝트의 경우 6개의 주 철골(하중의 대부분을 지지), 19개의 부 철골, 23개의 멀리언, 12열의 트랜섬으로 구성된다. Geometry composition process was a naming process introduced in order to manage all the structural parts. For this project, 6 main steel frames, 19 sub steel frames, 23 mullions, and 12 transoms were included.

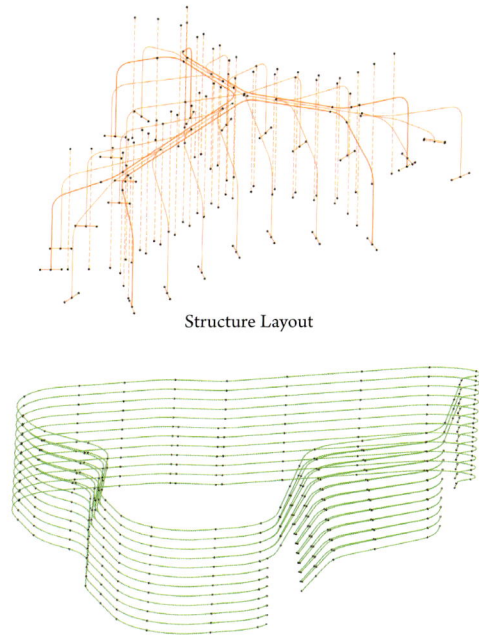
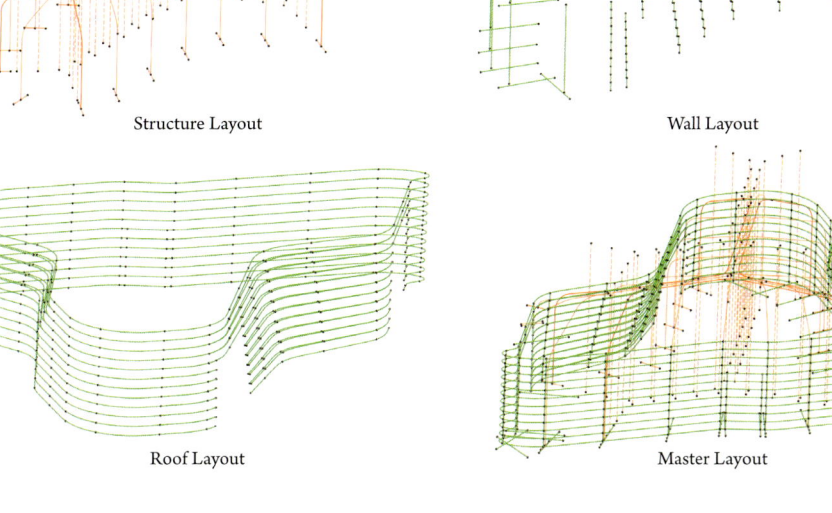

Structure Layout

Wall Layout

Roof Layout

Master Layout

fig. 4

glass had a functional purpose, it demonstrated a progressive technological advancement that would be 'the next generation in glass'.

## The NINE BRIDGES Pergola and Digital Technology

Before construction began, the NINE BRIDGES Pergola was proceeded in three different stages. First, a design method using a digital programme was extracted. Secondly, construction parts were measured by the 3D measuring device. Then lastly, the steel frame was manufactured and examined.

**Designing via CATIA**

As mentioned earlier, this project used the CATIA programme. Served Space (functional space for the users) and Servant Space (space for the structure and the building system) were combined, and functional parts were located as per the following building's shape. At this point, all the construction drawings had to match perfectly in order to minimize any error when on site. It began with analysing the 2D drawings, the stage at which the NINE BRIDGES Pergola's wall, roof, and structural data were contemplated and processed. Information taken from the finished 2D drawings was entered into Rhino software to complete the

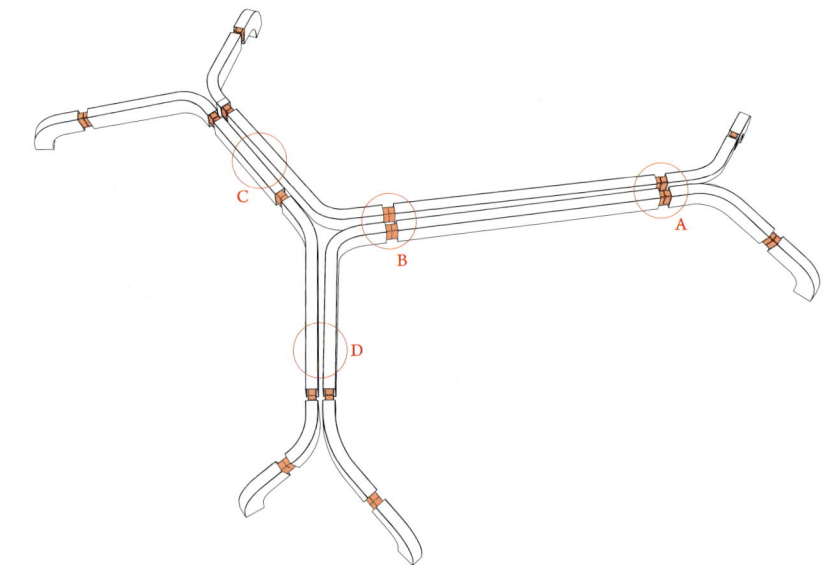

fig. 5

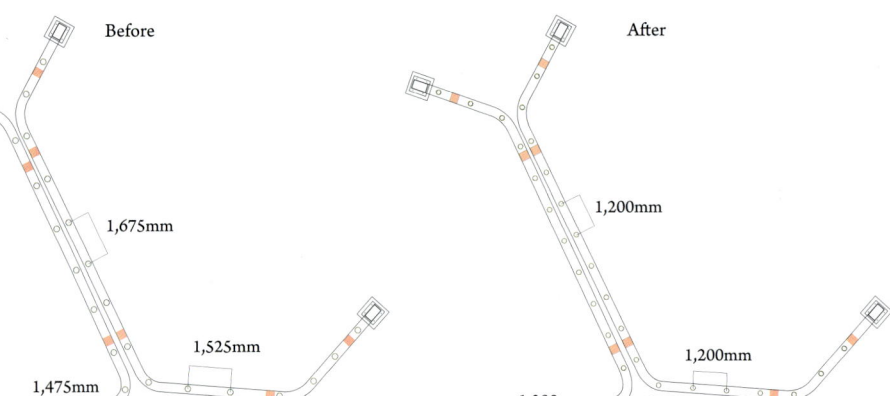

fig. 6

**fig. 5** 두꺼운 주 철골 안에 덕트를 철판 형태로 유지한 채 삽입하기 위해서 주 철골을 총 18개로 나눈다. In order to insert the duct, we maximized the portion of the linear steel parts and the separation distance is given. The main steel frame was divided into 18.

**fig. 6** 디퓨저의 배치는 MEP 도면을 작성해 기존의 도면과 비교해 역으로 수정하는 방법을 택했다. We decided to create MEP drawings, then to compare them with the original drawings to inversely modify them.

basic modeling, and then we used CATIA to finish the more complex details.

Throughout this process, MEP drawings had to be converted into 3-dimensional renderings. The design plan and manufacturing method of the ducts, which were the core elements of MEP, were also decided . Since ducts, serving as functional elements, were being inserted into the steel frames, which served as structural components, the method of connection between these core members could be determined. Also, for those construction parts that were being cut, the location of the cut to improve the efficiency of construction was also considered. The ways to modify the error, if any occured while construction modules were being transported and assembled on site, were also brought into consideration. When we were completing the actual construction on site, the modeling was continuously modified based upon a construction method with a higher efficiency rating, and then the 2D drawings were inversely modified. Even though the drawing analysis and modification had been completed, the construction components were in a state of complete disorder. In order to ease the pressures of the modeling stage and to situate the right construction parts in the right place when actually assembling them, certain rules were made to re-organize and re-position the components. If the result of the modeling at this stage in process was used to check the approximate location and shape of the building, then the modeling created by the next procedures was impeccable enough to be used for the actual production.

Considered as the core of digital design, a geometry composition process was carried out. This was a naming process introduced in order to manage all the structural parts. For this project, 6 main steel frames (supporting most of the weight), 19 sub steel frames, 23 mullions, and 12 transoms were included. The name of this structure was the 'Master Wireframe', which literally means the inflexible framework, and it became the standard for construction components that were added later. This was a very important line of work: by going through a period of standardization, the location of the soon-to-be-added structure and covering were automatically determined, and details could be added or modified by accessing the results of the data in real time.

This project required a very precise design plan. Even the manufacturing and construction methods were discussed throughout the design stage. Once the basic structural work was finished, the modeling for those construction elements acting as the skin of the building could begin. This included the data for the final form of the construction elements, as well as their finishing

자를 경우, 절단하면서 곡률의 변형이 생길 수도 있기 때문에 300mm 이격 거리를 주었다. 이러한 규칙으로 생산된 주 철골을 총 18개로 나누었으며, 단일 부재의 꼭지점과 꼭지점 간 최대 길이는 7,000mm였다.

주 철골 내에 삽입된 덕트를 통해 흐르는 공기를 밖으로 배출하는 디퓨저의 배치도 이번 프로젝트의 연구 과제였다. 기존의 MEP 도면에 표기된 개구율과 구조 산정 값을 바탕으로 개구부 간의 거리 1,200mm가 정해졌고, 2D CAD 도면에 작성된 내용을 바탕으로 3D 모델링 파일에 투영해보았다. 그러나 평면상에 표기된 개구부 간의 거리를 3D 도면상에 배치했을 때, 간격이 훨씬 커졌고 이는 전체 개구율에 영향을 주어 공기의 흐름에 영향을 미쳤다. 또한 모든 개구부가 같은 간격으로 배치된 것이 아니기에 미관상으로도 좋지 않다는 것을 모델링을 통해 발견했다. 해결책으로 모델링상에서 공기 조화와 관련된 MEP 도면을 새로 작성해 기존의 도면과 비교해 역으로 수정하는 방법을 택했고, 모든 개구부의 간격을 1,200mm의 간격으로 맞추었으며, 개수도 40개에서 48개로 늘렸다. 또한 불규칙적인 디퓨저의 간격을 조절하여 같은 간격을 유지하면서 3D 형상에서의 대칭을 만들었다. 이 과정을 통해 비정형 설계에서 설계 도면과 MEP 도면의 미세한 오차는 전체 기능과 외관에 큰 영향을 준다는 것을 경험했다.

입면에서 층을 구분해주는 330여 개의 알루미늄 돌출 캡은 유리와 함께 직접적으로 사용자의 시선을 받는 곳이기에 오차 없는 곡면의 형성과 깔끔한 마감처리가 중요했다. 따라서 일체형으로 알루미늄 캡을 원하는 모양으로 압출해 하나의 형태로 제작하는 것이 최선의 방법이었으나, 일부 구간은 부재의 굴곡이 심해 압출 바를 구부려서 제작하게 되면 성분의 정밀도가 낮아지고 품질이 저하되기 때문에, 빌트업 방식으로 제작했다. A 타입의 경우 직선 부재이거나 1개의 곡선과 직선이 만나는 경우였으며, 11개의 층까지 모두 동일한 형태이기에 알루미늄 캡을 압출하여 일체형으로 제작이 가능했다. 그러나 B, C, D 타입의 경우 부재들을 아르곤을 용접한 후, 외관에서 보았을 때 일체형으로 보일 수 있도록 매끄럽게 가공했.

디테일을 확정한 다음 제작 데이터 산출을 위해 모델링의 완성도를 높였다. 이 과정을 통해 건물을 구성하는 유리와 철골 개별 부재에 대한 데이터 값을 입력해 개별 부재의 관리와 전체 형상의 변화를 쉽게 파악할 수 있었다.

유리는 형태에 따라 평유리, 2D 유리[3], 3D 유리[4]로 분류한다. 유리의 사양에 따라 제작 방식과 소요 시간에서 큰 차이를 보이기 때문에 생산이 쉬운 평유리와 2D 유리는 수치만으로도 제작이 가능하다. 하지만 3D 유리의 경우 꼭지점 간의 대각 길이와 서로 다른 반지름이 만나는 지점에

**fig. 7** 입면에서 층을 구분해주는 330여 개의 알루미늄 돌출 캡은 오차 없는 곡면의 형성과 깔끔한 마감처리가 중요했다. 총 4개의 타입으로 구성되었다.

As 330 of the aluminum caps—which differentiate on an individual level across elevations—received the user's attention along with the glass, the curved surfaces must be formed and finished cleanly to an error-free standard.

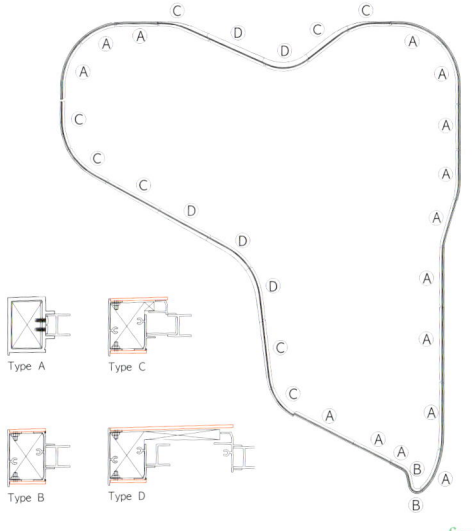

fig. 7

and assembly methods. It needed to be as perfect as possible so that it could be used directly in the field of manufacture and construction. We selected the Built-Up² method to minimize the on-site construction period – from building the structure to finalizing it by placing finishing materials – and tried to minimize the cuts necessary for the construction members.

The NINE BRIDGES Pergola was the first design to insert the building system inside the structure. In order to place the 2mm duct that remained as an iron plate inside the 12mm-thick steel frame, it was necessary to research the method of connection and the corresponding technology. Since a linear steel part is easy to insert, we maximized the portion of the linear steel parts and determined the location of the joint of the framework to be 300mm away from the point where the linear line and circular line intersected. The 300mm of separation was to avoid any deformation from happening when the endpoint of the circular line was cut off. Following this rule, the main steel frame was divided into 18, and the maximum distance between the vertices of individual members was 7,000mm.

Another research subject in this project was the location of the diffusers. Diffusers discharge external air via ducts inserted into the main steel frame. Based on the opening ratio rules written on the original MEP drawing and the computation result of the structure, the distance between the openings was worked out (1,200mm). Then information drawn on the 2D CAD drawings was used to project the distance between the openings on the 3D modeling file. However, when we actually located it on the 3D drawings, the distance became much wider – affecting the overall opening ratio, and eventually the air flow. Also, via modeling we were able to understand that since all openings were not evenly spaced they did not look good when visualised on the exterior of the building.

As a solution, we decided to create new MEP drawings for air conditioning during the modeling, then to compare them with the original drawings to inversely modify them. We also standardized

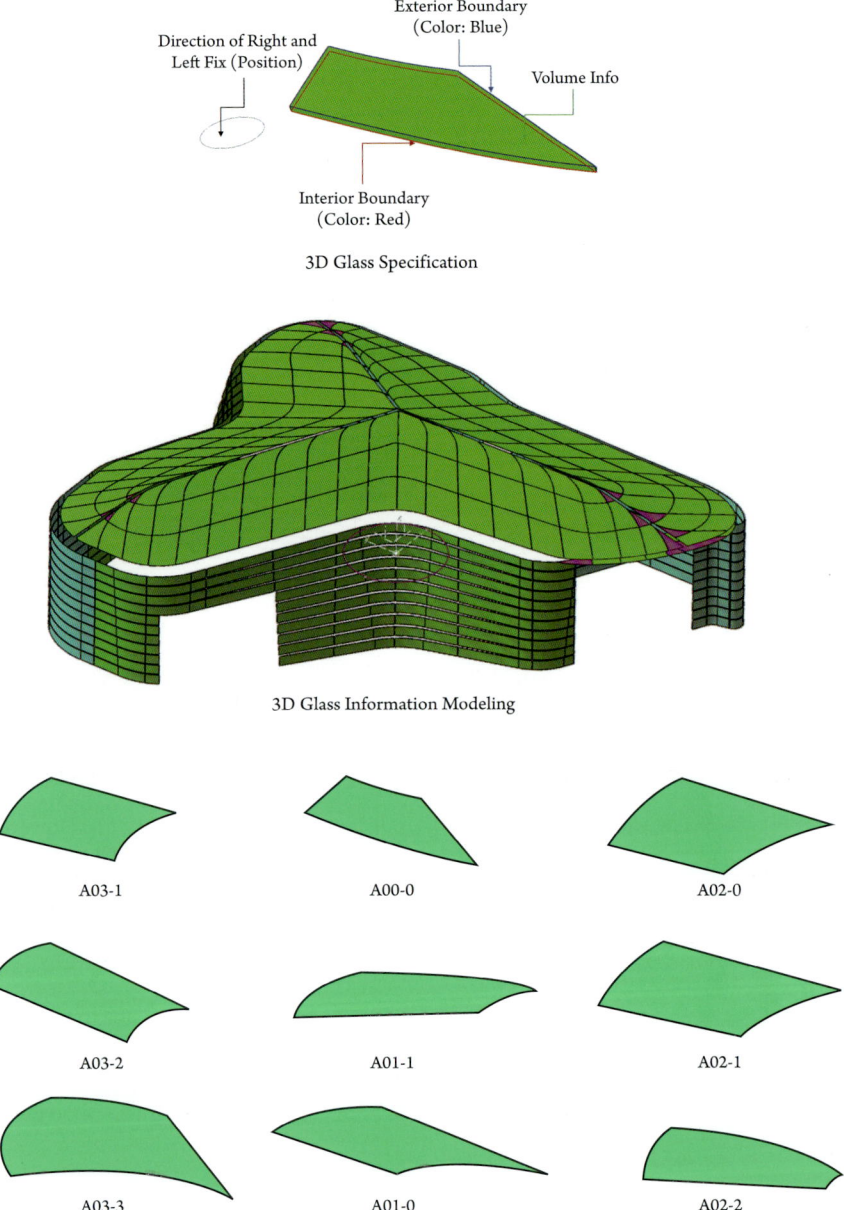

fig. 8

fig. 8 3D 프로그램을 통해 지정된 유리는 2차원에서 파악할 수 있는 네 번의 길이, 면적의 값을 산출한 다음, 그 산출된 값들과 함께 엑셀 값으로 저장된다.

Via the 3D programme, we were able to figure out the inside, outside, and the installation direction of the glass panels. For the glass parts, for which locations had already been determined, their lengths of four sides and their areas—which can be identified in the 2-dimensional—were calculated.

the distance between openings to be 1,200mm and increased the number of the opening from 40 to 48. Then we adjusted the uneven spacing between diffusers to keep the same distance and maintained the symmetry of the 3D form. By going through this process, we noted that for irregularly shaped building designs a very minute margin of error exists between the design drawings and MEP drawings, which significantly affects the function and the exterior of the building.

As 330 of the aluminum caps—which differentiate on an individual level across elevations—received the user's attention along with the glass, the curved surfaces must be formed and finished cleanly to an error-free standard. So the method of pressing aluminum caps into an all-in-one design was the best solution, but the problem was that a few of the parts were extremely curvy. In this case, bending the pressured aluminum bar would lower the degree of precision and quality of the elements and degrade the overall standard. It was therefore manufactured as a Built-Up type. In the case of type A, it was where a linear member was laid out, or a curvy member and a linear member were meeting. Since all glazings from the ground to the top shared the identical form, we were able to manufacture the aluminum caps as an all-in-one design. However, for type B, C, and D, construction members were welded with argon and smoothly treated to be seen as an all-in-one design.

Once the details were set, the grade of completion of the modeling was more developed so as to be able to extract data for manufacturing. Throughout this process, the resulting values of individual glass and steel parts composing the building were entered, in order to conveniently recognise how each construction member was managed and if there would be any deformation in the building.

The glass is classified by shape – a flat glass, a 2D glass[3], and a 3D glass[4]. Since the manufacturing methods and the time required vary greatly depending on the type of glass, the flat glass and the 2D glass—which are relatively easy to produce—can be manufactured with the acquired measurements. However, in the case of the 3D glass, more precise data was required such as the length of a line segment drawn between vertices and the meeting point of different radii.

Later on, via the 3D programme, we were able to figure out the inside, outside, and the installation direction of the glass panels. For the glass parts, for which locations had already been determined, their lengths of four sides and their areas—which can be identified in the 2-dimensional—were calculated. Later on, the name of each glass panel was identified and

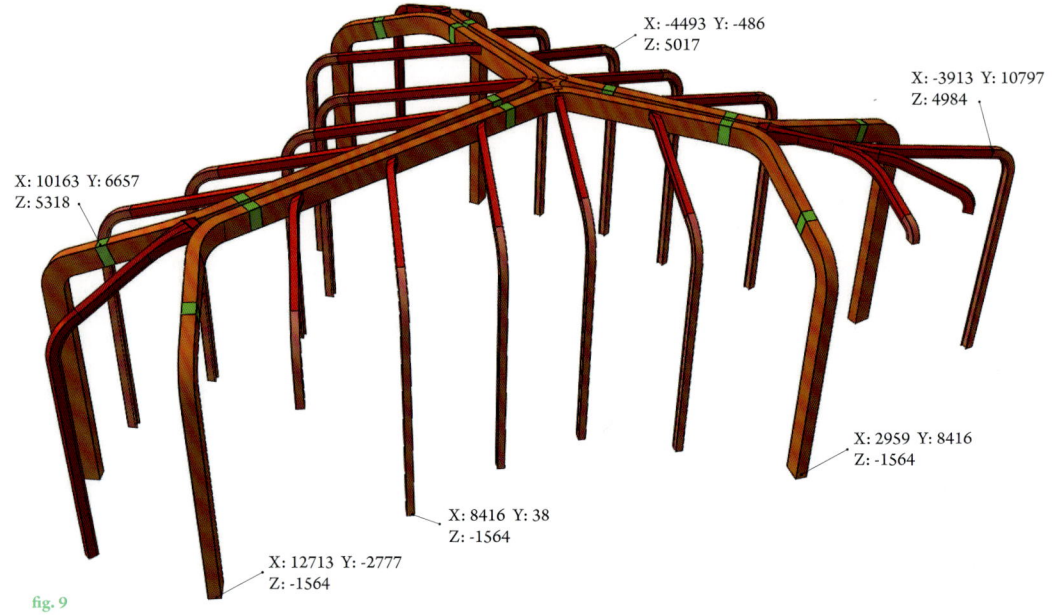

fig. 9

대한 정보가 필요하기 때문에 정밀한 데이터 값을 요구했다.

이후 3D 프로그램을 통해 유리의 내외부 및 설치 방향을 파악할 수 있도록 방향성을 지정해주고, 지정된 유리는 2차원에서 파악할 수 있는 네 변의 길이, 면적의 값을 산출한 다음, 최종적으로 이름이 지정되어 배치도에 표시가 되며, 산출된 값들과 함께 엑셀 값으로 저장된다.

유리와 마찬가지로 벽체와 철골구조도 이름을 붙여 관리한다. 모델링을 바탕으로 6개의 주 철골, 19개의 부 철골은 각각의 위치 및 형상에 대한 정보를 얻고, 일련의 순서로 이름을 매겨 배치도 및 전개도를 산출한다. 마찬가지로 지붕 철골도 일련의 순서를 만들어 이름을 지정한다.

기존의 기술이 모델링을 통해 복잡한 형태를 시각적으로 파악하기 쉽게 구현하고 이를 통해 모형 제작의 수준에 그쳤다면, 나인브릿지 파고라는 카티아를 통해 모든 부재의 위치적 정보를 정밀하게 파악하고 실제 제작이 가능한 도면을 추출하여 제품 생산을 완료했다는 것이다. 구조적 성능뿐만 아니라 부분별로 다른 부재의 짜임새를 만들어, 풍부한 디테일을 드러냈다. 부재의 제작 방식과 재료의 표현 방식, 그리고 나인브릿지 파고라의 안정적인 구조는 우리의 한 단계 올라선 디지털 설계의 가능성과 제작 기술의 진보를 보여주었다.

fig. 9 6개의 주 철골, 19개의 부 철골은 일련의 순서로 이름을 매겨 전개도를 산출한다.
The locations and shapes of each of the 6 main steel frames and 19 sub steel frames were acquired, converted into a series of names, and computed with the site plan and elevations.
fig. 10 철골을 현장에 설치하기 전, 철골이 콘크리트 슬라브와 만나는 지점에 대한 정확한 측량이 필요했다.
Before installing the steel columns on site, the location – where steel frames were meeting the concrete slabs – needed to be measured accurately.

displayed on the site plan, and the following data was saved in Excel according to the calculated values.

In the same manner, we managed the walls and steel structures by naming them. Based on the modeling, the locations and shapes of each of the 6 main steel frames and 19 sub steel frames were acquired, converted into a series of names, and computed with the site plan and elevations. Likewise, roof steel frames were classified according to the series of order and named.

The original technology was restrictive, as we only able to realise a complex form that could be easily visualised via modeling, and only then we were able to pass onto the production stage. The potency of the NINE BRIDGES Pergola was that, by using CATIA, the locations of all individual elements could be precisely figured out, and drawings for manufacturing were automatically drawn to finalise production. It also displayed ample details regarding not only the structure but also the organised partial connections between different construction parts. The NINE BRIDGES Pergola's method of manufacturing construction parts, the demonstration of materials, and the stabilized structure elevated our potential for digital design and development in manufacturing technology.

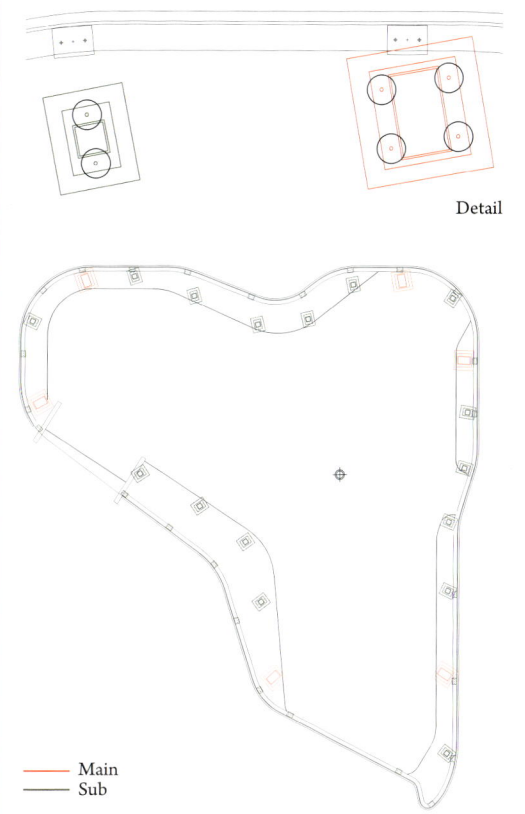

Detail

Main
Sub

fig. 10

**Measuring the Location of the Steel Columns (3D measuring device)**

Before installing the steel columns on site, the location—where 6 main steel frames and 16 sub steel frames were meeting the concrete slabs—needed to be measured accurately. Therefore the location of the 22 members was measured by the 3D measuring device at the factory. The distance between the points was measured and checked inversely with the dimension of the drawings. The

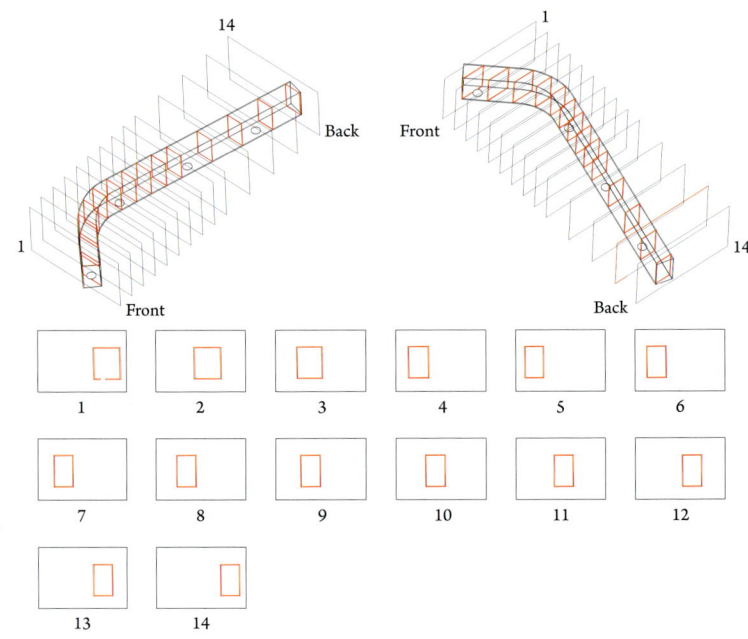

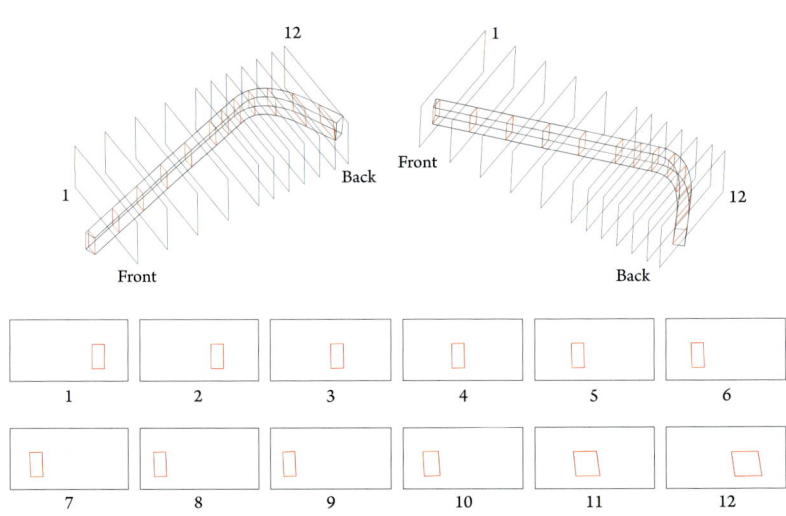

fig. 11

fig. 11 3차원 형상의 철골을 제작하기 위해선 보조 부재인 지그가 필요하다. In order to make the 3D shape for the steel frame, a subsidiary part to adjust the curvature of the members was required. known as 'Zig'

fig. 12 회색 면이 원본 모델링 데이터이며 노란색 선이 검측한 결과를 나타낸 것이다.
The grey surface was the original modeling and the yellow lines were the result of the examination.

strength of the 3D measuring device was that, as long as the machine did not move, the measuring could continue until it matched the dimensions of the drawing.

The method of measuring was as follows: after installing the 3D measuring device, the 'work point'—the central point of the measurement—was determined. This work point indicated the locational standard of the measuring device, and also became the standard point for the location of the 22 later-to-be-displayed points. The points, being measured, were the central points of the anchor bolt which held the base of the column together, and the numbers of points, being measured for the main steel frame and for the sub steel frame, were 4 and 2 respectively. After finishing the measurements, the measured distance between the points was checked to verify if they matched with that of the original drawings.

**Manufacturing and Examining the Steel Frame (3D Scanner)**

In order to make the 3-dimensional shape for the steel frame, a subsidiary part to adjust the curvature of the members was required, known as 'Zig'[5]. By cutting the steel frame structure from a certain distance and extracting the sectional information, the identical sectional plate was made. At this time, the manufactured plate was

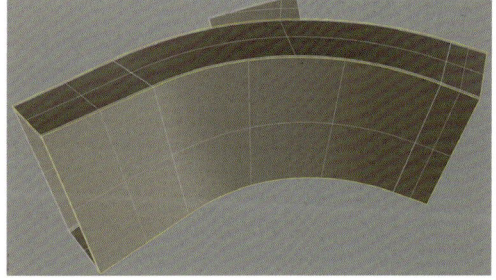
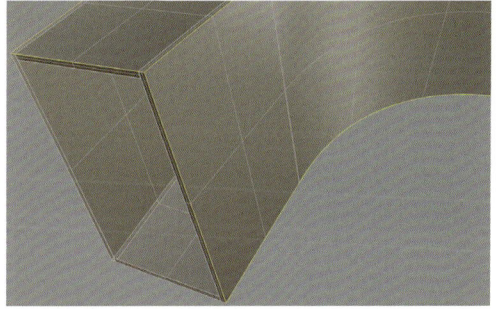
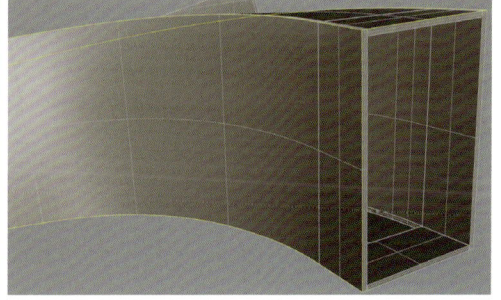

fig. 12

inserted into the steel frame inversely, to see if the curvature of the specific region was correctly applied. Where the bending was severe, the Zig was arranged with a greater density. Where the bending was gentle, the Zig was arranged widely.

To bend the steel frame with the Zig, the surface had to be heated. For this procedure, the heat needed to be adjusted according to the degree

**철골 기둥 지점 측량(3D 측량기)**

철골을 현장에 설치하기 전, 6개의 주 철골과 16개의 부 철골이 콘크리트 슬라브와 만나는 지점에 대한 정확한 측량이 필요했다. 따라서 공장에서 22개 부재에 대한 지점을 3D 측량기로 측량한 후, 지점 간 거리를 측정해, 역으로 도면상의 치수와 일치하는지 확인했다. 3D 측량기는 기계가 움직이지 않는 한, 정확한 도면상의 치수 값이 나올 때까지 지속적으로 측량이 가능하다는 장점이 있다.

측량 방법은 3D 측량기를 설치한 후, 측량의 중심이 되는 '기준점work point'을 지정한다. 이 기준점은 측량기의 위치적 기준을 나타내는 중심으로, 후에 표시될 22개 지점의 거리 값의 기준이 된다. 측량하는 점들은 기둥의 베이스 판을 고정하는 앵커 볼트의 중심점으로, 주 철골의 경우 4개, 부 철골의 경우 2개점을 측량한다. 측량을 완료한 후, 점들 사이의 거리를 측정해, 원본 도면의 거리 값과 일치하는지 확인한다.

**철골 프레임의 생산 및 검측(3D 스캐너)**

3차원 형상의 철골을 제작하기 위해선 부재의 곡률을 조절할 수 있는 보조 부재가 필요한데 지그[5]가 이에 해당한다. 모델링상의 철골 구조체를 일정 간격으로 잘라서, 단면 정보를 추출해낸 후, 그 단면과 동일한 형상의 판을 제작한다. 이때 역으로 제작한 판을 철골에 끼워서 특정 부분의 곡률 값이 올바르게 반영되었는지 확인할 수 있다. 휨의 정도가 심한 부분에는 지그를 촘촘하게 배치하고, 완만한 부분에는 넓게 배치한다. 지그와 더불어 철골을 휘게 하기 위해서는 면에 열을 가하게 되는데, 이때 면의 굴곡진 정도에 따라 가하는 열을 세밀하게 조절하는 것이 필요하다. 곡률이 심한 면에는 열을 집중적으로 가하고, 완만한 면에는 열을 적게 가해 곡률을 조절할 수 있다.

한치의 오차도 없이 설치해야 하므로, 개별 부재에 대한 꼼꼼한 스캔 작업이 필요했다. 제작이 끝나면 22개의 모든 구조체들에 대한 3D 스캔 작업을 시작했다. 이후에 철골 위에 부가적으로 설치될 부재의 기준이 되는 부분으로 3D 스캐너로 부재의 곡률 및 전체 치수의 오차를 측정하는데, 기본적인 검측 원리는 다음과 같다.

철골 표면에 측정 기기의 장착된 펜으로 점을 찍는다. 곡률이 큰 지점은 점을 촘촘히 찍고, 평면에 가까운 부분은 간격을 넓게 점을 찍어 완료하면, 그 점들로 면을 생성해 결과 값을 3D 파일로 추출해낸다. 그 결과 값을 카티아를 통해 원본 데이터에 입혀 오차의 정도를 판단해 오차가 큰 부재를 선별하고자 했다. 측정한 결과, 22개의 모든 부재는 3mm 오차 내로 제작이 되었으며, 곡률도 원본 데이터와 일치해 조립 가능한 부재로 판별되어 다시 제작해야 할 필요가 없었고 제작 시간을 최소화할 수 있었다.

또한 이번 프로젝트에서는 3차원 곡면의 형태를 가진 철골을 제작하기 위해서 부재의 휨 정도를 조절할 수 있는 기술이 필요했다. 곡면에서 오차를 3mm 이내로 만들기 위해 여러

of bending. Where the curvature was severe, the heat was applied intensively. Where the curvature is gentle, less heat was applied.

The installation did not allow for any error: individual members needed to be scanned precisely. Once the manufacturing was finished, all 22 structures were 3D scanned. Later on, the curvatures of the components and the margin of error for the overall dimensions were measured by the 3D scanner, which became the standards for the members to be added to the steel frame later. The basic examination principle was as follows.

First, print the points on the surface of the steel frame with the equipped pen of the device. Print the points at the parts of an extreme curvature with a greater density, and more widely at the nearly flat parts. The surface formed by these points was extracted from the 3D file. Then, by using the CATIA, the resulting values were applied to the original data. The margin of error was calculated, and then the construction member with the largest margin of error was clarified. After conducting this measurement, we found all of the 22 members that had been manufactured within 3mm of the margin of error, and the curvature also matched with the original data. All framing members were agreed as eligible for assembly without reproduction – minimizing the amount of time for manufacture.

Moreover, in this project a new technological development to control the constituent levels of bending became indispensable, in order to make a 3-dimensional curvy shape for the steel frames. To minimize the curvy surface's margin of error to within 3mm, we had to go through several procedures for production. The member with the largest degree of bending was selected for examination by the scanner and to be compared with the original data.

**Assembling and Examining the Steel Frame Structure (3D Laser Scan)**

The method of assembling the steel frame structure and conducting a 3D laser scanning process was as follows:

> Tentatively assemble the steel frame structure → Scan and examine the overall shape using the 3D laser scanner → Input data and optimize it → generate and edit the Point Cloud data[6] → Generate the Mesh data[7] → Edit the Point Cloud → Generate the Surface[8] modeling → Compare the result with the original modeling data → Modify the assembled structure → Transport it to the site.

In the case of the steel frame structure, even if the individual framing parts were carefully manufactured, the errors incurred during the

fig. 13

차례 제작 과정을 거쳐야 했다. 구조체 중 휨이 가장 심한 부재를 선정해 스캐너로 검측하여 원본 데이터와 비교했다. 실제로 이 부재의 오차는 1mm 이내였다. 이는 경험으로 축적된 우리만의 방법으로 어떤 형태의 철골이든 원본 모델링과의 오차를 최대한 줄여 생산할 수 있다.

**철골구조체 조립 및 검측(3D 레이저 스캔)**
철골구조체의 조립 단계 및 3D 레이저 스캔 과정은 다음과 같다.

철골구조체 가조립 → 3D 레이저 스캐너로 전체 형상 스캔 및 검측 → 데이터 입력 및 최적화 → 점 집합 데이터[6] 생성 및 편집 → 메시 데이터[7] 생성 → 점 집합 편집 → 표면 모델링[8] 생성 → 원본 모델링 데이터와 비교 → 조립된 구조체 수정 → 현장 운반

철골구조체의 경우 개별 부재가 정밀하게 생산되었다 하더라도, 시공상에서 부재 간 결합 과정에서 발생하는 오차가 전체 형상에 큰 변화를 줄 수 있기 때문에 공장에서 가조립을 진행했다. 대부분의 이형 부재는 지붕에 분포되어 있기 때문에, 지붕 조립 작업을 완료한 후 공장에서 1차 스캔 작업을 진행했다.

먼저 3D 레이저 스캐너를 이용하여, 대상에 대한 형상 정보를 빠르고 정확하게 측정한다. 이를 통해 역 설계 및 분석 등에 활용할 수 있다. 가조립된 형상의 스캔 작업을 통해 실제 형태와의 오차를 분석하고, 모델링과 상이한 부분을 사전에 확인해 현장에서 조립 시 오차를 최소화했다.

나인브릿지 파고라에서는 규모가 큰 작업에 사용되는 광대역 스캐너가 사용되었다. 대상을 기준으로 360도를 돌며 주변과 내부를 수차례 이동하여, 3차원 점 집합 데이터point cloud data를

**fig. 13** 공장에서 구조체 가조립을 완료한 다음 3D 레이저 스캐너를 이용하여 스캔작업을 한 결과를 원본 모델링과 병합한 것이다.
After the structure was assembled at the factory, the scan results were merged with the original modeling.
**fig. 14** 모델링과 상이한 부분을 확인해 현장 조립 시 오차를 최소화했다.
The margin of error compared to the actual shape was analysed. The difference was found in this advanced manner and minimized the error prior to assembly on site.

1. Mesh Model    2. Error Check    3. Suface Model    4. Output

fig. 14

process of connecting these parts could result in a major deformation. Therefore the tentative assembly was conducted at the factory. Also, most of the complex shapes were distributed on the roof. So once the roof was finally assembled the first scans were conducted at the factory.

First, by using the 3D laser scanner, we measured information concerning the shape of the building as quickly and as precisely as possible. The values obtained from this procedure could be repurposed for the inverse design and analysis. By performing the scanning to tentatively assemble the shape, the margin of error compared to the actual shape was analysed. The difference was found in this advanced manner and minimized the error prior to assembly on site.

A broadband scanner—which is a typical device for a large-scale project—was used in the NINE BRIDGES Pergola. The scanner rotated 360 degrees around its surrounding area as well as the interior of the structure multiple times, and would then read the 3-dimensional Point Cloud Data. If the obtained data was insufficient, it could make an impact on the data handling procedure and the accuracy of the final model. Hence, in order to obtain the most reliable data on the site, the geographic features were accounted for and a high tripod was used to conduct the scanning process

읽어냈다. 얻어낸 데이터가 미흡한 경우, 데이터 처리 과정 및 최종 모델의 정확도에 영향을 주기 때문에, 현장에서 정확한 데이터 취득을 위해 지형 지물이나, 고소 삼각대를 이용해 여러 위치에서 스캔 작업을 진행했다.

스캔 작업이 이루어진 뒤에는 초기 스캔 데이터들을 하나로 합치는 병합 과정이 먼저 진행된다. 무수히 많은 점으로 형성된 점 집합point cloud 데이터를 생성한 후, 불필요한 점들을 제거하는 작업을 거쳤다. 이 후 점들을 연결해 메시 데이터를 생성하여 그물망과 같은 형태로 변환한다. 이 작업은 표면 모델링을 위한 최초의 데이터가 된다. 아직까지는 하나의 매끄러운 면이 아니라 울퉁불퉁하고 가독성이 떨어져 완벽한 정보는 아니다. 이후, 메시mesh 데이터에서 원형에 최대한 가까운 모양으로 표면surface 모델링을 진행한다. 이 과정에서 오차를 최대한 줄여야 하기에 마지막으로 카티아에서 편집할 수 있는 파일로 추출하여, 원본 모델링 데이터와 비교 작업을 한다.

## 철골 시공

공장에서 가조립을 진행한 후 내린 결과, 부재의 조립에 걸리는 시간이 오래 걸리므로 공사 기간을 단축하기 위해서는 현장 조립을 최소화해야 하고, 전체 뼈대를 최대한 적게 분할해야 한다는 것이었다. 또한 3차원 형상의 부재들은 접합 방식도 일정하지 않아 시공 난이도가 높기 때문에 공사 기간을 고려하여 지붕을 먼저 조립한 후 인양하여 기둥에 접합하는 방법으로 시공을 진행하기로 했다. 클럽하우스라는 특수한 현장 상황을 고려해서, 부피가 큰 기둥을 포함한 구조체를 배치하는 방법의 연구도 필요했다. 철골 구조체의 설치 순서는 다음과 같았다.

철골 부재 해체 및 운반 → 지붕 구조체 조립 → 22개 기둥 지점 측량 및 설치 → 지붕 인양 및 용접 → 지붕 및 벽체 트랜섬 및 멀리언 설치 → 전체 스캔 및 수정

22개 기둥 지점의 측량은 공장에서 진행했던 것과 동일한 방법으로 3D 측량기를 통해 진행했다. 이후, 가조립 과정과 동일하게 지붕 구조체를 조립한 후, 크레인으로 인양해 기둥과 접합했다.

설치 완료 후 현장에서 3차 스캔 작업을 진행했다. 스캔 작업은 총 2대의 기계를 사용했으며, 장비로 인해 제한이 되는 부분과 시야를 가리는 부분은 지형 지물을 활용해, 옥상에서 스캔 작업을 실시했다. 최종 스캔 작업이 끝난 후, 원본 모델링 데이터와 비교했다.

가장 민감한 부위인 부재와 부재가 만나는

**fig. 15** 원본 모델링과 스캔 결과를 병합한 그림을 보면, 가장 민감한 부위는 부재와 부재가 만나는 지점, 부재의 곡률이 급격히 변하는 지점이다.

When we looked at and combined the drawing from the results of the scanning and of the original modeling, we found that the black surfaces were not aligned to that of the original. The most susceptible parts were those in which one framing part met another, or where the curvature of the parts radically changed.

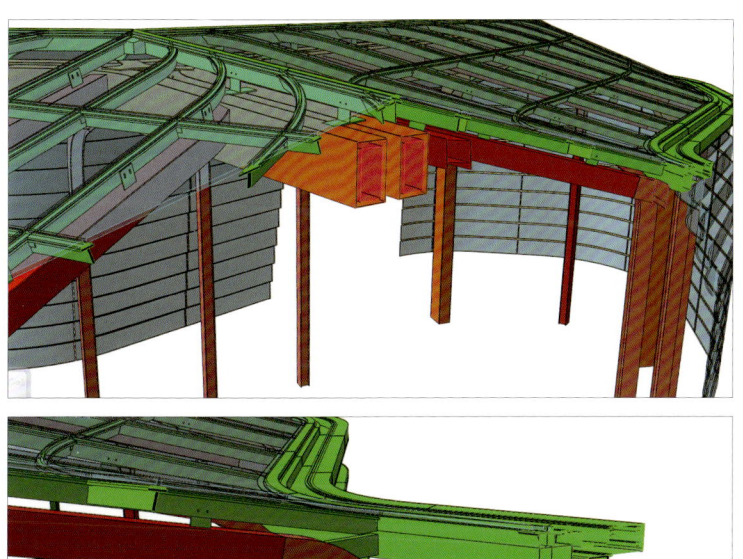

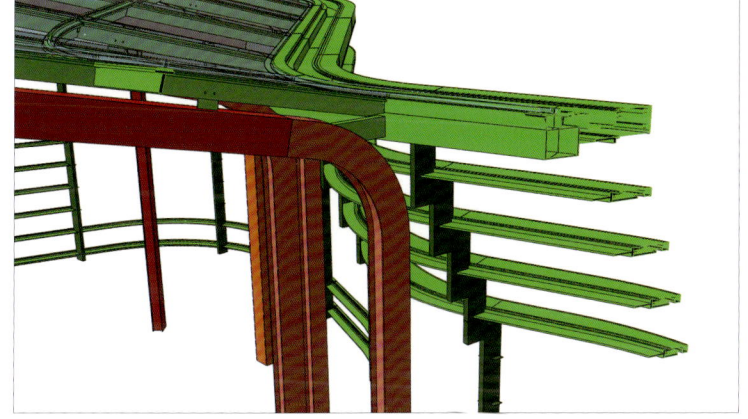

fig. 15

throughout many of the locations.

Once the scanning process was finished, another procedure – to combine the initial scanned data into a single group – was conducted. After the point cloud data was generated, composed of countless dots, the unnecessary dots were removed. Then the surviving dots were connected to generate the Mesh data and transformed into a net-like shape. This process became the first data set for the surface modeling. So far this is not a single smooth surface, but rather an uneven or poorly readable one of incomplete information. Yet, based on the Mesh data the surface modeling was modified to become as close to the original shape as possible. The margin of error had to be minimized during this process, so the CATIA-editable file was extracted, at last, to be compared with the original modeling data.

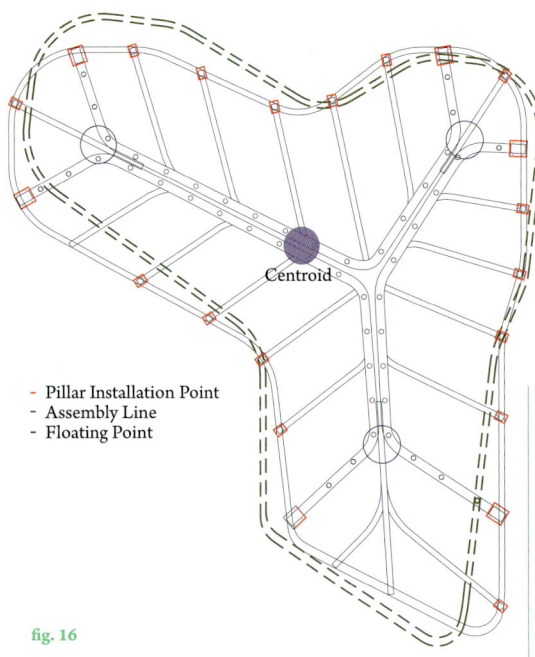

fig. 16

하나의 띠처럼 보이도록 마감처리를 했다.

## 유리 제작 및 시공

유리를 성형하는 기술은 다양한 방법이 있지만 대표적으로 핫벤딩[9] 방식과 콜드벤딩[10] 방식이 있다. 핫벤딩의 경우 유리를 높은 온도에서 성형하는 방식으로 가장 오래되고 많이 사용되는 방식이며, 콜드벤딩의 경우 최근에 개발된 방식으로 현장 조립 단계에서 유리를 성형하는 방법이다. 두 방법을 통해 유리의 다양한 곡률 값과 정밀함을 조절할 수 있으며, 건물의 구조와 복잡성에 따라 선택하는 방식도 바뀐다.

일반적으로 2개 이상의 곡률을 가지거나, 곡률의 값이 매우 작을 경우에 유리 성형 과정에는 높은 열을 필요로 하는 작업이 수반된다. 따라서 나인브릿지 파고라에서는 골프장이라는 특수한 환경 조건과 날카로운 형태의 정밀한 유리 생산을 위해 콜드벤딩 기술 대신 핫벤딩 기술을 사용했다. 2개 이상의 곡률 값을 가진 비정형 유리 생산 기술의 핵심은 '유리의 시각 왜곡 없이 얼마나 휘게 할 수 있을까'라는 것이다. 먼저 2D 유리의 경우, 위에서 설명했듯이 원통형 유리cylindrical glass라고도 부르는데, 이는 하나의 반지름 값을 가지기 때문에 중심축을 기준으로 대칭인 형태로 만들 수 있다.

이런 단순한 형태 외에 포물선 형태나 구

지점, 부재의 곡률이 급격히 변하는 지점을 분석했다. 그 결과, 모든 지점에서 오차범위 5mm 이내에 들어와 유리 설치에 문제가 없다는 결과를 얻었고, 마감재를 설치할 준비가 완료되었다.

마감재 중 가장 비율이 높은 알루미늄 돌출 캡, 지붕 마감 패널과 스크린, 그리고 디퓨저 및 기타 부재의 설치 작업을 시작했다. 알루미늄 캡 구간은 수량이 400개나 되고 제작 단계에서부터 마감처리를 깔끔하게 하여 운반했기에, 파손 없이 설치하는 것이 중요했다. 일체형으로 제작된 구간은 문제 없이 설치했으며, 조립으로 설치한 구간은 후에 추가 보강 작업을 진행하여 외관상

fig. 16 3차원 형상의 부재들은 접합 방식도 일정하지 않아 시공 난이도가 높기 때문에 공사 기간을 고려하여 지붕을 먼저 조립한 후 기둥에 접합하는 방법으로 시공을 진행하기로 했다.
Due to the inconsistent method of connecting the 3-dimensional shape of the framing members, construction seemed like it would be difficult to pull off. Therefore, construction proceeded by assembling the roof first and hoisted it later to attach to the columns.

**Constructing the Steel Frame**

After conducting a tentative assembly at the factory, we decided to reduce the amount of on-site assembly and split the overall structure into the least possible number of divisions. It was a treatment performed to shorten the construction period because assembling the constituent parts usually takes a long time. Moreover, due to the inconsistent method of connecting the three-dimensional shape of the framing members, construction seemed like it would be difficult to pull off. Therefore, construction proceeded by assembling the roof first and hoisted it later to attach to the columns. Considering the unique conditions of the club house, a method of locating the structure, including the bulky columns, also had to be studied. The method of assembling the steel frame structure was as follows:

> Dismantle and transport the steel frame structure → Assemble the roof structure → Measure the location for 22 columns and install them → Hoist the roof and weld together → Install the roof, wall transom, and mullions → Scan the overall shape and modify if necessary.

Measuring the location for 22 columns was completed with the 3D measuring device, in the same manner as it was in the factory. Afterwards, the roof was assembled similarly to the tentative assembling procedure and hoisted by the crane to attach it to the columns.

Following installation, the third scanning process was conducted on site. A total of two machines were used for the scanning process. The areas the machine was unable to measure or the areas obscured by a blocked view were measured by pinpointing the geographic features from the rooftop. Once the final scanning process was complete, the result was contrasted to the original modeling data.

When we looked at and combined the drawing from the results of the scanning and of the original modeling, we found that the black surfaces were not aligned to that of the original. The most susceptible parts were those in which one framing part met another, or where the curvature of the parts radically changed. By analysing these parts, the margin of error for all parts fell within 5mm. The glass was ready to be installed, as well as the finishing materials.

Of the finishing materials, mostly the used parts, such as aluminum cap, roof finishing panels, screens, and diffusers, were to be installed first. It was essential for the aluminum caps to be installed without incurring any damage: they numbered over 400 and their finishing was very neatly

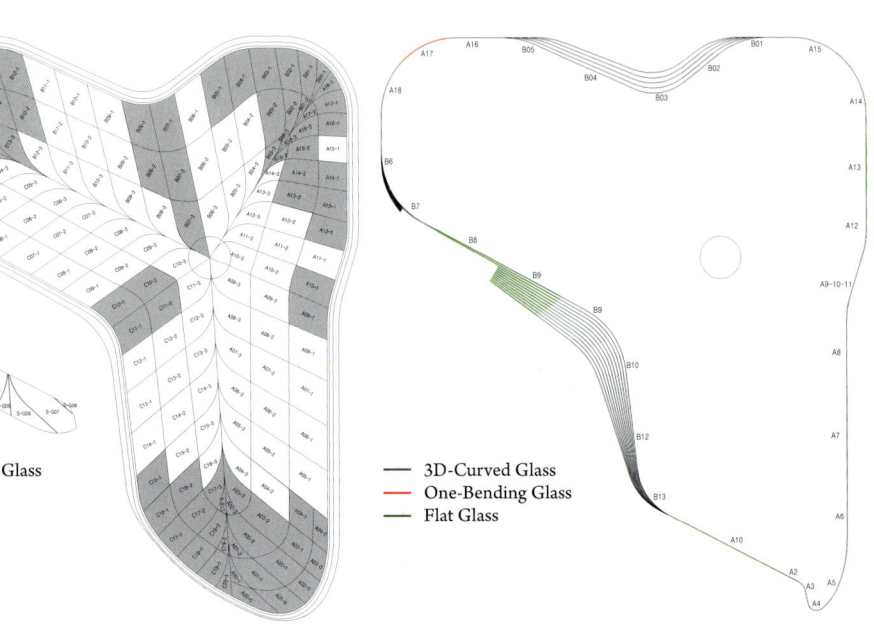

fig. 17

형태를 가진 3D 유리를 구형 유리spherical glass라고 부른다. 3D 유리는 두 종류가 있는데, 여러 개의 반지름 값을 가진 유리가 연결된 벽체 유리가 이에 해당한다. 그리고 반지름 값으로만은 완벽한 데이터의 표현이 불가능한 비정형의 지붕 유리가 있다. 이 유리를 완벽히 제작하기 위해서는 꼭지점 간의 거리 및 유리의 깊이를 비롯한 정밀한 데이터 산출이 요구되며, 성형 과정에서 특정한 축을 따라 단면을 형성해 철골 생산 과정에 쓰인 공법과 비슷한 틀이 제작되어 형태를 잡는다.

제작할 수 있는 최소 반지름의 크기는 유리 원판의 두께와 상관 관계가 있는데, 현재 기술로는 두께 4~6mm 원판의 경우 최소 100mm, 두께 10mm 원판의 경우 300mm의 크기를 생산해낼 수 있다. 또한 통상적인 곡면 유리를 최대로 정밀하게 만들었을 때 생기는 오차는 3mm 이내이다.

이 프로젝트에 사용된 지붕 유리는 6Low iron (Super-SE-2)#2 HS + 12Ar + 5Low iron HS / 1.52PVB/5Low iron HS이다. 벽체 유리는 6Low iron (Super-SE-2)#2 HS + 12Ar + 6Low iron HS로 사용했다. 지붕 유리는 총 173장으로 이 중 3D 비정형 유리가 90장, 평

fig. 17 나인브릿지 파고라에 사용된 지붕 유리는 총 173장으로 이 중 3D 비정형 유리가 90장, 평 유리가 83장이었으며, 벽체의 경우 총 310장 중 3D 비정형 유리가 226장, 2D 유리가 27장, 평 유리가 57장으로 3D 유리의 비중이 굉장히 높았다. The total number of roof glass panes was 173: 90 of these were 3D irregularly shaped glass panes and 83 were the flat glass panes. The total number in the wall assembly was 310: 226 were the 3D irregularly shaped glass panes, 27 were the 2D glass panes, and 57 were the flat glass panes.

done, after which they were transported from the manufacturing stage with the utmost care. Segments that were produced as part of the all-in-one design were installed without any problems. For the parts that were installed by the Built-Up, they went through an additional reinforcement process in order to be seen as a single band strip.

**Manufacturing and Erecting the Glass**

There are many ways to shape the glass, but the two prominent methods are Hot-Bending[9] and Cold-Bending[10]. In the case of the Hot-Bending, the glass is formed at a very high temperature. This is the longest and most frequently used method. On the other hand, the Cold-Bending was developed recently, which allows for forming the glass at the assembly stage on the site. By using these two methods, the curvature values and details of the glass can be adjusted. Depending on the structure and the complexity of the building, the method best-suited to these factors was chosen.

In general, when forming glass with either more than two curvatures or with a very small value of curvature, a higher temperature is required. Considering the fact that the NINE BRIDGES Pergola is a golf course, and in order to produce sharper glass in a more precise manner, Hot-Bending instead of Cold-Bending was used. The main idea behind the manufacturing technology of multi-curved glass with more than two curvature values was 'how much can we bend the glass without any optical distortion'.

In the case of the 2D glass—also referred as a Cylindrical Glass—has only one radius value as explained earlier, and it is possible to make a symmetrical shape based on a central axis. Other than such a simple form, a 3D glass structure with a parabola or spherical shape is called a Spherical Glass. There are two types of the 3D glass, and the curtain wall system, which is formed of an array of glasses with multiple radius values, falls under this category. There is also a roof glass which is irregularly shaped, and for which data cannot be perfectly presented with radial values only. In order to manufacture this glass perfectly the precise calculation of the data, such as the distance between vertices and the depth of the glass, is required, and during the forming process the mould – which is manufactured in a similar manner as the steel manufacturing process, where a section was built up by following a particular axis – is used to fix the shape.

The minimum radius value eligible to be manufactured depends on the thickness of the original glass. As far as the current technology

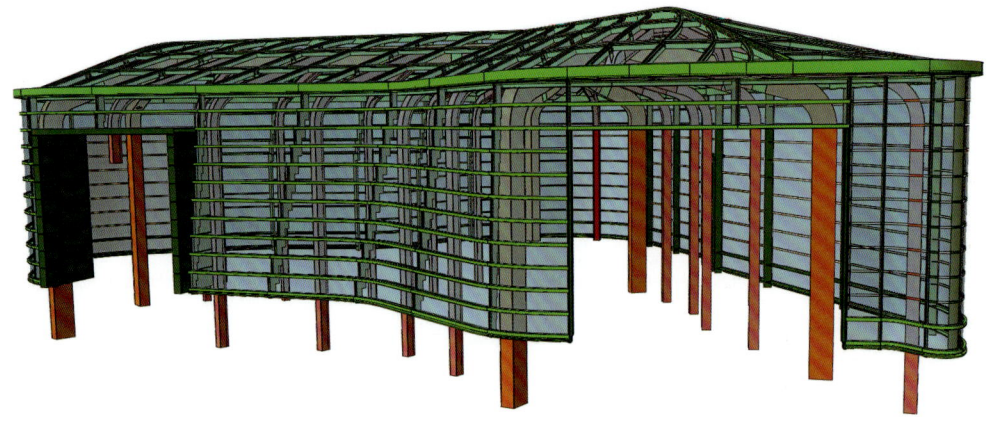

fig. 18

유리가 83장이었으며, 벽체의 경우 총 310장 중 3D 비정형 유리가 226장, 2D 유리가 27장, 평 유리가 57장으로 3D 유리의 비중이 굉장히 높았다.

복층 접합 유리를 구성하는 원판의 크기는 5, 6mm 원판으로 제일 반지름이 작은 부재의 값은 350mm 였다. 현재의 기술로 제작까지는 문제가 없으나 강화 과정 중, 유리 부재 중 날카로운 부분이 많고 급격히 곡률이 변하는 지점이 여러 번 파손되어 공정 방법의 변화가 필요했다. 이에 비정형 유리 강화 프레스기의 정밀도를 높이는 작업을 두 번 한 후, 모든 유리의 강화 과정을 마칠 수 있었다.

유리가 제대로 제작되었는지 확인하기 위해, 중국 공장에 철골을 검측할 때 사용했던 3D 스캐너를 가져가 검측을 진행했다. 생산된 비정형 유리를 모두 검측한 결과, 모두 1mm 오차 범위 내에 들어와 사용에 적합하다는 판정을 받았다.

오차 없이 설치된 뼈대를 바탕으로, 도면과 동일하게 만들어진 마감재가 설치되고, 그에 따라 유리는 제작 오류도 없이 동일한 위치에 딱 맞는 형태로 설치할 수 있었다.

fig. 18 나인브릿지 파고라의 강점은 어느 곳에서나 분해하고 이동 후 조립하여도 완벽한 조립이 가능하다는 정교함에 있다.

The strength of the NINE BRIDGES Pergola was its ability to be perfectly reassembled and relocated on any site after being dismantled.

concerned, the minimum of 100mm for the original glass depth of 4 – 6mm, and 300mm for the original glass depth of 10mm can be made. When any typical curved glass surface was manufactured as precisely as possible, the margin of error is within 3mm. The roof glass used for this project was 6Low iron (Super-SE-2)#2 HS + 12Ar + 5Low iron HS / 1.52PVB/5Low iron HS. For the wall, 6Low iron (Super-SE-2)#2 HS + 12Ar + 6Low iron HS were used. The total number of roof glass panes was 173:90 of these were 3D irregularly shaped glass panes and 83 were the flat glass panes. The total number in the wall assembly was 310:226 were the 3D irregularly shaped glass panes, 27 were the 2D glass panes, and 57 were the flat glass panes – the allocation of 3D glass was the largest.

The size of the original shape which forms the laminated tempered glass was 5 – 6 mm, and the glass components of the smallest radius value were 350mm. There has been no problem thus far in manufacturing the glass with the present technology. However, during the reinforcement process, the sharp edges and parts with a rapidly changing curvature were damaged multiple times, and so a change needed to be made to the manufacturing process. After carrying out a process twice to improve the precision of the multi-curved glass reinforcing press device, we were able to finish the rest of the glass reinforcing process.

In order to ensure whether the glass was properly manufactured, we brought the 3D scanner—which was used for examining the steel frame at the factory in China—and administered the examination. After testing every piece of irregularly shaped manufactured glass, all were verified to be within 1mm of the margin of error, and were proven eligible for use.

Considering the structural frame was installed without error, the finishing materials—which were produced along with the drawing—were installed, and the glass could be located at identical positions, without any manufacturing errors.
Drawn by ILJIN Unisco

## 주

**1 MEP**
건축설비 전반에 관여된 정보가 포함된 도면을 뜻한다. 기계, 전기 소방 등 건축물의 설비 전체를 의미한다.

**2 조립**
기계공학, 건축 토목 용어로 '조립한다' 라는 의미를 담고 있다. 여기에서는 공장에서 조립하여 완성된 형태로 가져와 설치하기 위한 밑작업으로 해석한다.

**3 2D 유리**
원통형 유리(cylindrical glass)라고도 부르며, 하나의 반지름 값을 가진 유리를 말한다.

**4 3D 유리**
구형 유리(spherical glass)라고도 부르며, 다양한 반지름 값을 포함하고 더 복잡한 기하학적 형태를 갖는 유리를 말한다.

**5 지그**
사전적 의미로 급격한 방향 전환 및 변경이라는 뜻으로, 여기에서는 급격한 방향 전환이 되는 지점들의 형상을 정확하게 만들기 위한 '틀'의 개념으로 이해한다.

**6 점 집합 데이터**
3차원 점 집합 데이터란 뜻으로, 3D 레이저 스캐너로 스캔해 추출한 직후의 결과 값을 말한다. 레이저로 측정 대상에 무수히 많은 점을 찍어 하나의 점 집합을 만든 후, 건물의 대략적인 형상을 형성하는 과정으로 인위적인 수정 작업을 거치기 전 결과 값을 의미한다.

**7 메시 데이터**
3차원 점들을 정리한 후 그 점들을 연결해 그물망으로 만들어 면을 형성하기 전 임시 형태를 만들어내는 단계이다. 이 과정을 통해 점 집합으로 이루어진 형태에서 보다 더 부드러운 형태로 바뀐다.

**8 표면 모델링**
메시 데이터로부터 면의 요철을 없애 하나의 부드러운 면을 형성하는 과정이다. 메시 데이터로부터 하나의 면을 형성할 때, 면을 구성하는 점 중 최고점과 최저점을 연결하여 만든다.

**9 핫벤딩**
성형 틀 위에 평유리를 얹은 후 원하는 곡률이 나올 때까지 600~650℃에 이를 때까지 열을 가해 유리 자중으로 성형 틀에 안착시켜 모양을 만드는 방법이다. 이 온도가 되면 유리는 점소성 유체 상태로 변하여 부드러워지기 때문에 원하는 형태로 주조할 수 있다. 유리의 형태[원 벤딩(one-bending) 또는 멀티 벤딩(multi-bending)]와 곡면 반지름의 크기에 따라 만드는 과정이 달라진다.

**10 콜드벤딩**
평유리에 외부 압력을 가해 원하는 모양으로 구부리는 과정을 말한다. 공기와 온도의 조작으로 유리를 구부리는 것이 아니기 때문에 건설 현장에 직접 운반해 원하는 모양으로 형성할 수 있다는 장점이 있다. 그러나 현재까지 이 방법으로 곡률을 조절할 수 있는 기술의 한계는 곡면의 반지름 값이 1만mm 이상인 유리만 생산할 수 있다는 점이다. 따라서 나인브릿지 파고라 현장에는 더 세밀한 기술을 요하는 방법이 필요했다.

**Footnote**

**1  MEP**

An acronym for Mechanical, Electrical, and Plumbing. Indicating an architectural drawing that contains information about the overall building system, such as mechanical, electrical, fire protection and so on.

**2  Built-Up**

A mechanical, architectural and civil engineering term for 'to assemble'. In this context, it implies the base work for assembling the construction members at the factory, transporting the finished structure to the site and installing it.

**3  2D Glass**

A glass with a single radial value; also referred to as a Cylindrical Glass.

**4  3D Glass**

A glass with various radial values and of a more complex geometric shape. Also referred to as a Spherical Glass.

**5  Zig (Casting)**

By definition, it is a sudden turn and change in direction. In this context, it is understood as a concept of 'casting' to make an accurate formation of points when a sudden turn occurs.

**6  Point Cloud Data**

A data group of 3-dimensional points, which is the result extracted immediately after scanning the subject of measurement by the 3D laser scanner. It is a process of forming an estimated shape of the building, by marking innumerable dots on the subject and combining them into a single group of points. It is also the resulting value – before going through any artificial modification process.

**7  Mesh Data**

Indicates a stage of forming a temporary shape. 3-dimensional points are sorted out and connected together to create a mesh, prior to creating a surface. Through this process, the shape becomes smoother than the one formed by the group of points.

**8  Surface Modeling**

A process of removing surficial prominence and depression from the Mesh data to form a single smooth surface. When creating a single surface based on the Mesh data, the maximum and minimum points composing the surface are connected.

**9  Hot-Bending**

A process of placing a sheet of flat glass on the mould and heating it to 600 – 650℃ until the glass self-deforms into the mould. Once it reaches the high temperature, the glass becomes a viscoplastic fluid: a smooth condition which is also easy to be cast into the desired shape. The process of manufacturing the glass may vary depending on the shape of the glass (either One-Bending or Multi-Bending) and the size of the radius of the curved surface.

**10  Cold-Bending**

Applying exterior pressure on a sheet of flat glass to bend it into the desired form. This method does not require air or temperature adjustment, so the advantage is that the glass could be transported and shaped as desired on site. However, with this method, a curved glass surface with a radius of more than 10,000mm can only be manufactured – due to the current limitations in the technology of adjusting the curvature. Therefore, the Pergola of the Club at Nine Bridges was in need of a glass manufacturing method requiring more detailed technology.

CJ 건설

## 현장의 기록
**Field Note**

CJ E&C

**2016. 10. 07**
<u>3D 구조물 제작에 대한 고민</u>

국내에 외장마감이 3D로 구성된 건축물은 다수 있으나 유리로만 형성된 경우는 드물어서 국내의 기술력에 대한 검증이 우선되었다.
또한 2017년 10월 19일에서 22일까지 제주에서 개최되는 PGA TOUR THE CJ CUP @ NINE BRIDGES의 성공적인 진행을 위해서는 9월 초 공사를 완료하고 사전에 운영을 해봐야 했기에 공정별 일정에 대한 부담이 많았으며, 국내 유리 구조물 사례를 확인해본 결과 실패한 경우도 있었다. 현장에서 비주얼 디스플레이 목업[1]을 통한 검증도 시간적인 제약 때문에 현실적으로 불가능했다.
나인브릿지 파고라 철구조물은 국내 제작 공장에서의 가조립과 3D 스캔으로 품질 확인 및 즉각적인 조치가 가능했다. 하지만 이중 곡률 유리는 해외 제작에 의존할 수밖에 없었으며 국내로 반입하기 전에 품질을 검증하고 관리할 수 있는 시스템을 필요로 했다.

**2016. 12. 15**
<u>토공 및 기초공사</u>

지하 1층은 기계실로 EHP 공조기가 설치되고 외측 트렌치 내에는 배수용 배관을 설치하여 물고임을 방지했다. 구조물 자체가 이중 곡면이 유기적으로 결합된 디자인으로 평면 또한 곡률이 각기 다른 선형이 조합되었으며 골조 자체가 유리의 곡면과 일치해야 했기에 일반적으로 트랜싯[2]을 사용한 먹메김 작업[3]이 불가능했다. 그래서 주변에 있는 건물에 기준점을 설치하고 트림블 로보틱 토털 스테이션 장비[4]를 사용하여 곡선 구간은 200~400mm 간격으로 표시해 설계된 곡면과 일치시켰다. 사용 방법은 포인트점이 표현된 오토 캐드 도면을 트림블 태블릿 PC에 입력하여 현장 시공을 위한 레이아웃 포인트를 추출한 후 트림블 장비와 연동하여 작업점을 표시했다.

**2017. 01. 25**
<u>1층 골조공사</u>

나인브릿지 파고라는 나뭇잎의 잎맥(물관 등)을 통하여 영양분이 공급되듯이 설비와 구조가 일체화된 자연 자체의 공간을 구현하는 것이 설계 콘셉트였다. 급·배기용 덕트와 전기 등 공급해야 하는 인프라를 건물 외측 구간에 형성된 공동구를 통하여 메인 구조체(철골)에 연결했다.
공조 방식으로는 당초 40개의 천장형 디퓨저로 급기를 계획하였으나, 건물 규모에 적합한 부재 크기를 유지하기 위하여 천장형 48개와 바닥형 12개의 디퓨저로 최적화된 공조 시스템을 구성했다.

## 2016. 10. 07
### How will they realise the 3D structure

There are many buildings whose exterior finishing was done in 3D, but those that are completely covered in glass are rare. The verification of domestic technology was often prioritized.

PGA TOUR THE CJ CUP @ NINE BRIDGES was scheduled for Jeju in 19 – 22 October 2017. In order to ensure the successful hosting of this event, the construction stage was supposed to be finished by early September. The scheduling for each process was another onerous burden. It was impossible to perform the necessary verification by using a Visual Display Mock-Up[1] on-site, mainly due to the time constraints.

For the steel structure of the NINE BRIDGES Pergola, it was most feasible to assemble it tentatively at the domestic factory, to conduct a 3D scan, to run a quality verification test, and to take immediate action if required. However, it was impossible to manufacture 2-Way Curved glass in the domestic factory. There was no other option than relying upon a foreign manufacturer for production – which also required a special system to verify the quality of the glass.

## 2016. 12. 15
### The Earthwork and Foundation Work

The first basement floor was a mechanical room in which the EHP conditioning equipment was installed. In order to avoid water ponding, a pipe for a drainage system was built inside the exterior trench. The structure itself was a design of an organically combined, doubly curved surface. Since the framework had to match with this curved glazing surface, it was impossible to conduct a general Marking[2] work using Theodolite[3]. With this in mind, a benchmark was set in a building nearby, and the curvy section of the building form was indicated every 200 – 400mm using Trimble Robotic Total Station[4] equipment to match the original design of the curved surface. The method of use was as follows: an Auto CAD drawing was displayed through individual dots entered in the tablet PC to extract layout points for the site construction. Then, by connecting the PC to the Trimble equipment the work points can be indicated.

## 2017. 01. 25
### The Framework on the First Floor

As the nutrition of a leaf is supplied through its veins (or vessels), the design concept of the NINE BRIDGES Pergola was informed by a desire realise a natural space, in which the equipment and structure would be unified. The infrastructure to provide air supply and return ducts, electricity, and so on, was connected to the main structure (steel frame) via a HVAC system, which was built on the exterior.

This design was altered to account for the appropriate sizes of the structural component parts. This modified and optimized design featured 48 ceiling diffusers and 12 floor diffusers.

 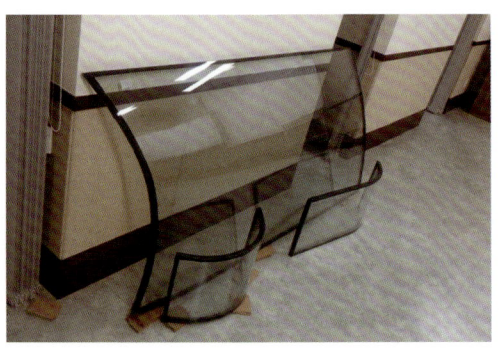

**2017. 03. 12**
**구조물 설계 / 제작 방법 검토**

일반적인 3D 건축물은 라이노Rhino를 사용하여 설계된다. 하지만 나인브릿지 파고라의 경우는 철골구조체 내부에 기계, 전기, 통신 등 건축물을 사용하고자 유지하기 위한 기능과 커튼월 마감 역할을 하는 요소들이 유기적으로 혼합되어 있는 구조물로 각각의 공정별로 필요로 하는 공간과 정밀성, 요구 성능이 달랐다.

　도면에서 표현되는 부재들이 현장에서 조립이 가능한지 사전 확인작업이 필요했기에 공정별로 수많은 부재들을 한 도면에 표현하며 검토하기로 했다. 이에 CAD 도면의 2D 정보를 라이노를 통하여 기본적인 3D 설계를 한 후에 카티아를 사용하여 불일치되는 부분을 구체화했다. 공정별로 제작 또는 구매 예정인 부재(부품)를 모델링하여 조립되는 순서대로 구현해 가면서 시공 효율성과 요구되는 성능이 유지되도록 설계를 완성했다. 이는 모형 제작이나 보여지기 위한 모델링이 아니라 자동차를 설계하듯이 실제 구현을 위한 작업이라는 것에 의미가 있다.

**2017. 03. 20**
**3D 유리 제작**

1면 곡유리는 국내시장에서 일정량의 수요가 있었지만, 이중 곡면 유리의 경우는 사용량이 경미하여 기술 및 생산 설비에 대한 개발이 적었다. 또한 초기 단계부터 원하는 크기의 이중 곡면 유리를 적당한 시기에 공급받기에는 어려울 것이라 판단했다. 국내에 설치된 3D 유리 사례로는 판교 미래에셋빌딩(중국제품)과 여의도 전경련회관(스페인 제품) 등이 있었다. 스페인 제품의 경우 제작기간(도면 확정 후 공장출하 기준 4개월)에 대한 부담이 커서 우선 중국 공장에 샘플 제작을 의뢰했다. 제작된 샘플이 현장에 반입되어 확인한 결과 치수 및 곡률이 나인브릿지 파고라의 기준에 적합함을 확인하여 발주·진행했다. 국내에서 생산된 철구조물의 제작 정밀도는 사전 확인이 가능하나, 유리의 중국 공장 자체 검측 방법은 대각 길이 및 곡률 깊이를 측정하는 것으로 실제 사면이 바에 부착되는 현장의 여건과는 부족한 점이 있다. 그래서 현장 담당자가 별도 장비인 프로라이너를 사용하여 중국 공장에서 검수를 진행하고 검증 완료된 제품을 국내로 반입하는 시스템을 구축했다.

2017. 03. 12

## Review of the Method of Design and Structural Production

In most cases, the design of 3D structures is performed using Rhino. In the case of The NINE BRIDGES Pergola, however, as a structure composed of various elements intended for the use and maintenance of the structure overall, such as machinery, electrical systems, communication networks, as well as the finishing of the curtain wall (all of which are integrated organically within the steel structure) the requirements regarding space, precision, and performance were different for each aspect.

As vital confirmation work on whether the key materials delineated in the floor plan could in fact be assembled on-site was required beforehand, a large number of these key materials were included in the floor plan according to their respective processes in order to to be properly examined. For this, after bringing the 2D information from the CAD floor plan to Rhino for a basic 3D design, the misaligned parts were specified by CATIA. Following the steps per process from the production stage, modeling, and assembly stage of key materials that are yet to be ordered, the design was completed to maintain construction efficiency and the performance requirements. This work is significant in that it is not merely miniature model production or visual modeling, but a move towards the actual manifestation of automotive design.

2017. 03. 20

## Manufacturing 3D Glass

There has been demand for 1-WAY CURVED glasson the domestic market, yet the application of 2-Way Curved glass has been low. As the less well-tested technological and production facility in the field, understanding the ideal sizes of 2-Way Curved glass at the most appropriate juncture in the project was always going to be difficult.

Some of the domestic projects that installed 3D glass were Pangyo Mirae Asset (glass made in China) and the Federation of Korean Industries (glass made in Spain). We requested a sample to be made at the factory in China first, because in the case of Spanish manufacturer the production period (4-months from the drawing confirmation to the factory shipment) was staggering long. Once the manufactured sample was brought into Korea, its measurements and curvatures were confirmed under the standards of the NINE BRIDGES Pergola, and the final order was placed. The degree of precision for domestic-produced steel structures could be checked in advance, and yet a new system to test the glass had to be established for the following reasons: the self-examination method of the factory in China was limited to measuring the diagonal length and the depth of the curvature, which was insufficient for a construction site in which the actual four sides of the glass were to be attached at the BAR. Therefore, our site manager executed an inspection at the factory in China by using special equipment, a Proliner. After this inspection, only the verified products were included in the project.

**2017. 04. 14**
**철골 샘플 및 공장 제작**

사각형의 철골구조재를 2차원 곡면으로 제작하는 것은 큰 어려움은 없으나 비정형 건물의 특성상 뼈대가 되는 철골구조물의 정밀도가 품질에 큰 비중을 차지하므로 비틀림이 들어간 3차원 구조물의 제작과 품질을 확인하는 방법에 많은 고민을 하게 되었다.
철골과 덕트의 제작 방법은 철판을 1차 상온 벤딩 후 열을 가하여 지그에 맞도록 곡률을 세밀하게 조절했으며, 덕트와 철골재 사이에 설치되는 단열재는 용접열에도 품질에 이상이 없는 세라믹보드(1,200℃~1,400℃ 열에 견딤)를 사용했다. 이렇게 제작된 샘플은 3D 레이저 스캔을 하여 기존 설계모델(3D)과의 오차를 검증한 후 본 제품 제작에 착수했다. 검측이 완료된 단일 부재는 생산 공장에서 가조립을 했으며(부재 간 연결은 볼팅접합[5] → 현장 조립 시에는 용접) 3D 스캔 후 10mm 이상 차이가 발생하는 구간은 보정 작업과 해체 및 도장(하도) 공정을 거쳐 현장에 반입했다.
가조립 방법은 트림블 장비를 사용하여 기둥 위치를 먹메김 후 가대(C형강 및 JACK으로 구성)를 설치하여 부재 높이를 조정했고 이후 주 빔을 조립하고 서브 빔을 조립하는 순서로 진행했다.

**2017. 04. 14**
**Steel Samples and Factory Production**

While there is not much difficulty in working a rectangular steel structural form into a 2D curved surface, however, because the precision of the steel structure acts as the main support it takes up a significant consideration of the unique irregularity of the structure. For the production of the steel and ducts, by heating the steel plate after bending it first at room temperature, the plate was precisely adjusted in its curvature ratio to fit the Zig, and ceramic boards that will not succumb to changes in quality at welding temperatures were used as the insulating material between the duct and the steel material. Then the examined single member was tentatively assembled at the production factory (structural members were connected by bolting[5] when assembled on site, welding was used) then the form was 3D scanned. For any area with more than 10mm of difference, it was revised, painted to be delivered on site. During the process of tentative assembly, the Trimble equipment was used to mark the location of the columns, and the frame was installed to adjust the height of the structural component parts.

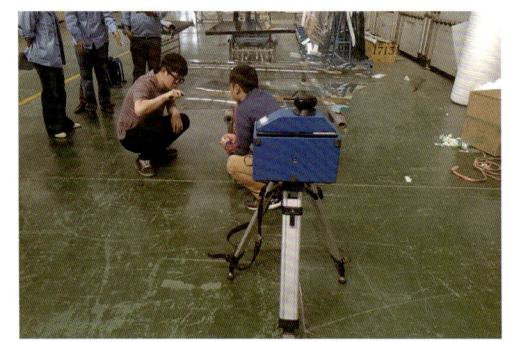 

**2017. 05. 11**
유리 제작 및 검수

**2017. 06. 01**
유리 반입 일정 협의

중국 공장과 계약상의 문제점으로는 품질 기준과 파손 시 책임 여부에 관한 내용은 명시되어 있으나 납기일에 대한 클레임과 보상은 명확하지 않았다. 이에 문제 발생을 사전에 인지하고 즉각적인 조치가 이루어지도록 생산 날짜에 공장을 방문하여 검측을 통과한 제품을 납품하는 것으로 진행했다. 1차 검수는 2016년 4월 17일~18일에 진행되었으며 제작된 이중 곡면 유리의 크기 및 곡률을 측정했다. 측정 기기는 프로라이너 장비로 검측의 주요 목적은 유리 곡률과 전체 형상이 당초 모델링과 동일하게 나왔는지 확인하는 것이었다. 측정 방법은 장비에 연결된 펜으로 유리의 표면에 점을 찍고 그 점들을 연결해 하나의 평면을 만들어 기존의 모델링 파일과 겹쳐서 오차를 확인했다. 검측한 곡면 유리는 A02-1과 A03-2(유리의 모양과 위치가 중요하므로 각 유리에 번호를 매겼다.)로, 곡률이 심한 부재였으나 원본 모델링과 결과값을 비교했을 때 코너 부위 오차 1mm, 곡률 오차 0.04mm 이내로 당사의 기준에 적합함을 확인하고 후속 공정 진행했다.

2017년 5월 11일, 제작 현황 및 납기일정을 확인했다. 62% 가량 제작 완료된 상태로 순차적 반입 시 현장 공사 완료일인 2017년 7월 31일을 준수하는 무난한 일정이었다. 하지만 이후 제작되는 3D 유리의 반강화 과정에서 파손이 발생(유리의 예각 부분이 많고 곡률이 크다는 이유)하여 유리 제작용 장치를 정비해야 한다는 것을 중국 측으로부터 통보받았다. PGA TOUR THE CJ CUP @ NINE BRIDGES의 개최를 준비하기 위해서는 기존 일정의 준수가 절실했다.
프레스의 정비는 품질 확보상 꼭 필요하다고 판단되었기에 '보수가 완료된 후 어떻게 해야 제주 현장의 제품을 우선적으로 제작할 수 있을까?'라고 고민했다. 결과적으로 '솔직하게 현재의 절박한 상황을 설명해주고 상호 신뢰를 바탕으로 접근해보자'라는 전략을 세우고 3차로 공장을 방문(2017년 6월 1일)했다. 물론 단기간 내에 서로의 마음을 알 수는 없었지만 '이번 프로젝트가 PGA 투어를 준비하기 위한 건물이라는 것'과 '클럽나인브릿지가 세계 100대 골프 코스 중 43위(2015년)인 클럽으로 여기에 중국의 제품이 설치되는 것은 큰 자랑이다'라는 것을 강조하여 장시간 설득 끝에 7월 1일 잔량을 출하하는 것으로 협의했다.

## 2017. 05. 11
### Manufacturing the Glass and Examining

We visited the factory on the date of manufacture and conducted an examination. Only the products that passed this assessment process were to be delivered. During the first inspection, held between 17 – 18 April 2016, the size and curvature of a single sheet of manufactured 2-Way Curved glass was measured. The main purpose of this examination was to verify whether the curvature and the overall shape of the glass was identical to the initial outlines at the modeling stage. The measurement was taken by the Proliner equipment. The method of measurement was as follows: by using the pen connected to the equipment, dots were placed on the surface of the glass and were connected to draw a plane. The plane was then created and overlapped with the original modeling file to check the margin of error. The types of curved glass that had passed the examination were A02-1 and A03-2, both with severe curvatures. However, once the measured result was compared with the measurement of the original modeling, the corner error and the curvature error were within 1mm and 0.04mm respectively. Since both products satisfied our standards, we progressed onto the next stage.

## 2017. 06. 01
### Consulting on the Delivery of the Glass

On 11 May 2017, the status of the manufacturing, as well as the projected date of delivery, was confirmed. At that time, the overall production was 62% complete. However, during the semi-reinforcement stage of the 3D glass manufacturing process, significant damage had been incurred (due to the large number of acute angles and the high curvature). We were notified by the factory in China that their press machine (for the 2-Way glass manufacturing) had to undergo vital maintenance work. Accordingly, a third visit to the factory (on 1 June 2017) was scheduled. Up until the visit, we struggled to find a way to receive the final products by our desired date. The maintenance of the press machine seemed necessary to ensure the highest possible quality of the product. Therefore, we strived to figure out a way of asking the factory in China – once the machine was repaired – to prioritize our project in Jeju above all others.

So our plan was to emphasise two facts: one was that 'this project was to prepare for the official tournament of PGA TOUR THE CJ CUP @ NINE BRIDGES', and the other was that 'it would be a great privilege to install a Chinese product at The Club at NINE BRIDGES, which was ranked at 43rd on the world's top 100 golf courses (2015)'. After many hours of persuasion at the factory in China, the discussion regarding the delivery of our shipment came to its conclusion – the final products would leave Tianjin on 1 July 2017.

**2017. 06. 05**
<u>구조물 현장조립</u>

조립상 주요 사항으로는 공장에서 가조립 후 3D 스캔과 보정을 통하여 품질이 확보된 구조물을 현장에서 동일하게 재조립하여 구현하는 것이었다.
일반적인 철골 조립 방법인 기둥 → 대들보 → 보 순서대로 시공하기에는 고소 작업을 위한 과다한 가설재의 투입과 조립 중인 부재를 고정하기 위한 별도의 시설물이 필요하였기에 공장에서 가조립 시 사용했던 가대를 동일하게 설치하여 시공 정밀도를 확보하고 고소 작업은 최소화하는 방향으로 시공 계획을 수립했다.
지붕 구조물은 기둥의 간섭을 피하기 위하여 방향을 틀어서 가조립을 했고, 기둥은 트림블 장비를 사용하여 기초면 상부에 4개 꼭지점을 표시한 후 앙카볼트 후매립 공법으로 시공했다.
조립된 지붕 구조물은 250톤의 메인 크레인과 25톤의 보조 크레인을 사용하여 야간에 인양하여 기둥과 체결 작업을 완료했다(12시간 소요).

**2017. 06. 05**
<u>The On-site Assembly of the Structure</u>

A major consideration in reassembling it was to build a structure identical to the temporary structure at the factory in China – which has been tentatively assembled, 3D scanned and modified to ensure the quality.
The usual sequence for a steel frame construction was to build the column → girder → beam. Instead, the construction plan was set as follows: the frame which was previously used at the factory—for the tentative assembly of the structure—was reinstalled on-site in the same manner. This marked an attempt to ensure the precision of reassembly while minimizing the higher placed work.
To avoid any interference with the column, the roof structure was temporarily assembled in a slightly shifted direction. For the construction of the column, four vertices were initially marked on the foundation using the trimble equipment, and then the column was anchor bolted to and encased by the foundation.
The assembled roof structure was hoisted by a 250-ton main crane and a 25-ton sub crane to be connected to the column (the whole procedure took 12 hours).

**2017. 06. 22**
**벽체 커튼월 조립 / 도장 / 유리 설치**

벽체 커튼월은 25톤 크레인 2대와 내부에 시저 리프트를 사용하여 조립했으며, 기둥의 설치 방법과 동일하게 트림블 장비로 기준점을 잡아 설치한 후 수평 T-바를 고정했다. 커튼월을 포함한 전체 철골 작업이 완료된 후 3D 레이저 스캔을 통한 검측 및 보정 작업과 시공 누락된 부위를 찾아 공사를 완료했다.
제주도의 특성상 바람이 세어 외각 쪽에 보양조치를 하였음에도 불구하고 외부로 비산되어 피해가 발생해 풍속이 3m/s 이하인 조건에만 도장 작업이 가능했다.
유리는 제작 후 납품 시에 넘버링을 하여 정 위치에 시공했다. 또한 지붕 유리 및 실란트 작업을 우선적으로 실시하여 내부 마감 작업과 벽체 구간 유리 설치를 병행했다.

**2017. 06. 22**
**Curtain Wall Assembly / Painting / Glass Installation**

The curtain wall was assembled using two of 25-ton cranes and a scissor lift. The method of assembly was similar to that of the main column: the trimble equipment picked the benchmark for installation, and then the horizontal T-BAR was fixed. Once the steel framework, including the curtain wall system, was complete, the examination and the modification were conducted by the 3D Laser Scanner. Any omitted work discovered later was also addressed and concluded. Since the construction site was its strong wind gusts, additional efforts to protect the painting had to be made at the exterior perimeter. Yet the paint was still scattered towards the outside, and subsequently sustained damage. As a result, the painting work was only possible when the wind blew at a speed lower than 3m/s.
The manufactured glass was numbered when shipped, so it could be built in the correct location. The work on the roof glass and the sealant was performed at an earlier point, and the interior finishing was carried out along with the curtain wall installation.

**2017. 07. 25**
**내부 마감 및 조경공사**

바닥 마감은 비닐 직조 카펫(폴리에스테르 필름), 콘크리트 폴리싱, 현무암 중 제주의 특성을 강조하기 위하여 현무암 패턴 깔기로 진행했다. 등기구는 라인형으로 철골구조물 속에 매립하여 설계 콘셉트를 반영했다.
내부 조경의 경우 곶자왈에 자생하는 지피식물로 구성하여 건물 내부에서도 제주 원시림을 느낄 수 있도록 표현했다.
외부 조경은 제주 산간지역에 자생하는 식물로 계획하여 자연스럽게 흐르는 숲 같은 정원을 조성했다. 외부 식재 사이로 보이는 건축물의 선형을 통하여 흥미를 유발시키되 내부 바닥 마감과 동일한 소재의 계단을 설치하여 물이 흐르는 것과 같이 자연스러운 진입 공간을 구성했다.

**2017. 07. 25**
**Interior Finishing and Landscape Design**

For floor finishing material, basalt rather than vinyl weaving carpet (polyester film) or concrete polishing was chosen to imply the uniqueness of the site in Jeju. The linear types of lighting fixture were embedded in the steel structure to reflect the overall design concept. Ground cover plants that naturally grow in the Gotjawal Forest were planted inside the structure, so that the interior atmosphere could resemble Jeju's native forest. For the landscape design around the perimeter of the site, plants that naturally grow in the hilly region of Jeju were laid out to create a naturally flowing, forest-like garden. The outline of the building, made visible between the plantings, could appeal to one's attention, while the stairwell, made with the same material as the building's interior flooring, has also been created to form a natural entry way that flows like water.

주

**1 비주얼 디스플레이 목업**
공사 전 건축물의 외벽과 구조체의 가장 중요한 부분을 실물 크기로 제작 현장 내에 설치하여 시간의 경과에 따른 품질을 확인하는 것이다.

**2 트랜싯(시어도라이트)**
각을 측정하는 측량기계로 공사를 위한 먹메김 작업 시에 사용한다.

**3 먹메김 작업**
골조공사 시 거푸집을 설치하기 위한 기준선 작업한다.

**4 트림블 로보틱 토털 스테이션 장비**
좌표를 입력하여 시공점을 찾는 방식에서 발전하여 CAD 도면을 입력한 후 필요한 포인트를 추출하여 작업점을 표시한다.(태블릿 PC에 프리즘과 추출한 포인트와의 거리가 표현됨)

**5 볼팅접합**
철골 부재 간 접합은 전체 용접접합이나 가조립 시에는 해체의 용의성과 부재 손상을 방지하기 위하여 볼팅접합으로 진행한다.

**Footnote**

**1  Visual Display Mock-Up**
Manufacturing a full-scale model of the most significant part of the building's exterior wall and the structure at the pre-construction phase. A quality performance test is also conducted over the passage of time.

**2  Marking**
Drawing a bench mark to set up the formwork for the frame construction

**3  Transit (theodolite)**
A survey machine to measure the angle. Used for marking before any construction starts.

**4  Trimble Robotic Total Station equipment**
Developed as the means for locating construction points after entering the coordinates. Once the CAD plan is input, the points required can be extracted to display the points of construction. (The distance between the prism and the extracted points is displayed on the table PC)

**5  Bolted connection**
Steel members are mostly welded, yet for a tentative assembly they are bolted to avoid the possibility of being dismantled or damaged.

**설계** (주)조호건축사사무소
**설계담당** 조준희, 홍봉귀, 정문영
**시공총괄** CJ 건설(주)
**시공총괄담당** 노민수
**철구조 / 커튼월 /유리** (주)일진유니스코
   - 총괄PM 정길수
   - 철구조 및 외장설계 황정서, 김창영
   - BIM 김덕준, 김선희
   - 시공 현장소장 이덕준
**구조설계** 터구조
**건축일반공사** 일호종합건설(주)
**기계·전기설계** 에이스엔지니어링
**조경** (주)뜰과숲
**감리** 아뜰리에 11
**디자인감리** (주)조호건축사사무소
**위치** 제주 서귀포시 안덕면 광평로 34-156
**용도** 클럽하우스
**대지면적** 481,238m$^2$
**건축면적** 264.60m$^2$
**연면적** 320.79m$^2$
**규모** 지상 1층, 지하 1층
**높이** 6.48m
**구조** 철골조
**외부마감** 로이복층유리, 제주현무암

**Architect** JOHO Architecture (Lee JeongHoon)

**Design team** Cho Junhee, Hong Bong-gwi, Jeong Moonyoung

**CM** CJ E&C

**CM team** Noh Minsu

**Steel structure / Curtain wall / Glass** ILJIN Unisco

- Project Manager Jeong Gil-soo

- Steel structure & Curtainwall Design Hwang Jeongseo, Kim Chang-young

- BIM Kim Dukjun, Kim Sun-hoi

- Site superintendent Lee Duk-jun

**Structural engineer** TEO Engineering

**Construction** ILHO construction

**Mechanical and electrical engineer** ACE Engineering

**Landscape design** Garden in Forest

**Construction supervision** Atelier 11

**Architectural supervision** JOHO Architecture

**Location** Seogwipo-si, Jeju-do, Korea

**Programme** Club House

**Site area** 481,238m$^2$

**Building area** 264.60m$^2$

**Gross floor area** 320.79m$^2$

**Building scope** B1, 1F

**Height** 6.48m

**Structure** Steel Frame Structure

**Exterior finishing** Low-e pair Glass, Jeju Basalt

**CJ 건설** / 클럽나인브릿지는 CJ 그룹 경영철학인 'OnlyOne'정신을 바탕으로 한라산 600고지에 조성된 자연친화적인 골프클럽이다. 2017년 US GOLF Magazine 선정 세계 100대 골프코스 41위(2017년), US PGA TOUR THE CJ CUP @ NINE BRIDGES 개최(2017년), US GOLF Digest 선정(미국제외) 세계 100대 골프코스 23위(2017년), 한국 10대 골프코스 7회 연속 1위라는 경이적인 기록을 이룩했다. 뿐만 아니라 US PGA Tour 및 US LPGA 등 권위 있는 메이저대회 개최와 세계 최초의 아마추어 골프대항전인 월드 클럽챔피언십(WCC)을 기획해 세계적인 골프대회로 육성시키면서 명실상부한 대한민국 대표 골프클럽으로 자리매김하게 되었다.

**일진 유니스코** / 일진유니스코는 1975년 건물외장사업을 시작하여 한국에 최초로 커튼월 공법을 도입했으며, 커튼월 업계를 선두하고 있는 회사이다. 당사는 다양한 디자인의 초고층 고난도 건물을 짧은 공기로 시공하는 것으로 알려져 있다. 최근 서울시청, 동대문 디자인 플라자, CJ R&D 센터, 전경련 회관 포디움과 같은 3D 건축물을 시공했다.

**조호건축** / 조호건축은 건축가 이정훈이 2009년에 서울에 설립한 건축사무소다. 이정훈은 성균관대학교에서 건축과 철학을 전공하고, 프랑스 낭시 건축학교에서 석사학위를, 파리 라빌레트 건축대학에서 D.P.L.G.를 취득했다. 조호건축은 재료를 일종의 기하 측정의 단위로 설정하고 이들의 군집과 가감을 통한 방법을 통하여 디자인을 발전시키고 있다. 이러한 재료의 측정은 재료가 지닌 의미를 대지가 지닌 컨텍스트 속에서 재해석하는 것을 의미하며 궁극적으로 건축공간을 인문학적 토대 위에 재구축함을 목표로 한다.

**CJ E&C** / The Club at NINE BRIDES set in natural surroundings under the gaze of Mt. Halla at an altitude of 600 meters is a natural friendly golf club based on the spirit of "OnlyOne" the management philosophy of the CJ Group. This prestigious golf course is ranked 41th best course in the World by US GOLF MAGAZINE in 2017, 23rd best golf course (excluding the USA) by US GOLF Digest, and has been Korea's #1 course for 7 consecutive years. By hosting prestigious competitions and tournaments, that include: US PGA Tour, US LPGA, and the World Club Championship first in the world, The Club at NINE BRIDGES has truly become the representative Golf Club in Korea.

**ILJIN Unisco** / ILJIN Unisco was in the facade business since 1975; being the first to introduce the Curtain Wall technology in Korea, this company has emerged as one of the leaders in the exterior construction industry. This company is known to have short construction durations, while implementing various architectural designs for complex high-rise buildings. In the past recent years, the company has participated in the construction of three dimensional structures, including buildings such as the Seoul City Hall, Dongdaemun Design Plaza, CJ R&D Center and the Federation of the Korean Industries Podium.

**JOHO Architecture** / JOHO Architecture is an architecture firm established in Seoul by architect Lee JeongHoon in 2009. Lee studied architecture and philosophy at Sungkyunkwan University before earning a master's degree in architectural materials at Ecole Nationale Supérieure d'Architecture de Nancy and an architects license (D.P.L.G.) at Ecole Nationale Supérieure d'Architecture de Paris-La Villette. In JOHO Architecture, architectural materials are set as a unit of 'geo_metry' and designs are developed through the grouping, addition and subtraction of these units. This 'material_metry' refers to the reinterpretation of a material's meaning within the context of a site.

**Pergola**
of The Club
NINE BRIDGES